NATURAL DISASTER RESEARCH, PREDICTION AND MITIGATION

AF848797

TSUNAMIS

ECONOMIC IMPACT, DISASTER MANAGEMENT AND FUTURE CHALLENGES

NATURAL DISASTER RESEARCH, PREDICTION AND MITIGATION

Additional books in this series can be found on Nova's website under the Series tab.

Additional e-books in this series can be found on Nova's website under the e-book tab.

NATURAL DISASTER RESEARCH, PREDICTION AND MITIGATION

TSUNAMIS

ECONOMIC IMPACT, DISASTER MANAGEMENT AND FUTURE CHALLENGES

TIANXING CAI
EDITOR

New York

For permission to use material from this book please contact us:
Telephone 631-231-7269; Fax 631-231-8175
Web Site: http://www.novapublishers.com

Library of Congress Cataloging-in-Publication Data

Tsunamis : economic impact, disaster management, and future challenges / editor, Tianxing Cai (Dan F. Smith Department of Chemical Engineering, Lamar University, TX, USA).
pages cm
Includes bibliographical references and index.
ISBN: 978-1-62808-682-9 (soft cover)
1. Shore protection. 2. Tsunami damage. I. Cai, Tianxing, editor of compilation.
TC330.T78 2013
363.34'94--dc23

2013029192

Published by Nova Science Publishers, Inc. † New York

CONTENTS

PREFACE

Two years ago, my research interest is mathematical modeling and process simulation in order to optimize the industrial operation, improve the manufacturing efficiency and increase the business profit. The objective functions in my optimization model will only be operation profits and process efficiency. The magnitude 9.0 earthquake and consequent tsunamis in Japan 2011 have upset my conventional wisdom about the mission of engineering. Never have I been aware enough of the fact that the tsunami is so close to all the aspects of our real life: communities, markets, economic systems, industrial operation, natural environment, medical care, emergency response, and the whole societies. The first priority of engineering design and optimization is to guarantee the operation safety and make a good preparedness of emergency response planning for the potential abnormal situations. The objective functions of my mathematical models have been shifted from the maximization of operation profit to the minimization of potential risk, environment impact or recovery time for emergency response. More proactive thinking has been put into my research to create a sustainable and safe operation environment and more potential risks caused by natural or anthropogenic activity have been taken into consideration in my research.

Therefore I was pleased to accept the tasks when I was invited by NOVA Publishers to write the book chapters and become the editor for this coming book "Tsunamis: Economic Impact, Disaster Management and Future Challenges ". This book is timely and in some ways timeless; the issues discussed within its pages are matters that are of interest to all people across the world and really across time.

The presented chapters focus on the theoretical positions and cognition about engineering thinking and suggestions for the tsunami in different segments of research fields.

The first chapter presents the innovative integration of geographic information system with point-of-care testing for emergency response and disaster management after tsunami. The 2004 Earthquake and Tsunami in Southeast Asia had devastating effects on coastal areas in Southern Thailand, where several thousand people lost their lives from drowning and associated injuries, while many more were displaced from their homes and businesses by destructive flooding, which disrupted roads and shut down transportation acutely. One of the hardest hit provinces was Phang Nga facing west to the Andaman Sea. After the acute Tsunami event, a damage assessment field survey of 4 provinces—Krabi, Phang Nga, Phuket, and Trang was conducted. Later Phang Nga Province was surveyed in detail to assess how healthcare facilities had recovered and adapted. Recently, a needs assessment of diagnostic testing was performed, including point-of-care (POC) testing, which can be defined as testing at or near the site of care and includes both self-testing and noninvasive monitoring. The chapter writers found significantly enhanced implementation of POC testing including increased use of noninvasive oxygen saturation monitoring in lieu of invasive in vitro blood gas testing. Currently, they are innovating terrain mapping and geographic information systems (GISs), which can be used to determine health centers, primary care units, community hospitals, and regional hospitals at risk for future tsunamis or other natural disasters. The GIS comprises a comprehensive database that includes all healthcare facilities in the province and also locates POC resources, backup, and preparedness factors, such as the critical distances from the most vulnerable population clusters along the coast and on islands to their sites of immediate, referral, and follow-up medical care. Elevation data in the GIS can be used to position heliports for future acute rescue and to determine critical land transportation paths for rescue. Overall, the conclusion can be reached small-world network of healthcare delivery in Phang Nga Province recovered nicely, but future leadership should adopt more advanced tools, such as GISs, to improve future preparedness and resilience.

The second chapter talks about the adaptive capacity building through public participation. The studied area has been identified as the first climate refugees on the west coast of Taiwan. Due to the groundwater over-pumping for the industry usage, the land subsidence has been observed during the past 20 years. With the global climate change, the phenomenon of sea level rising has worsened the problem of flooding in that area. Besides land subsidence and flooding, water shortage, water pollution, industrial decline, destroyed public intra-structure and the migration of young people to cities are also the critical problems in the area. Therefore, how to build the adaptive capacity of the residents to communicate with the government and corporations for enhancing the well being of the locals becomes a pivotal issue in the recent years. This project has started since 2006 and the action research approach has been used. Since 2006-2008, local major and influential stakeholders were interviewed for a better understanding of the problems people faced in that area. From 2009-2011, different types of community activities and classes have been implemented. The locals learned the related knowledge and express their own opinions through the above channels. In order to contact the most vulnerable people who are the poorest group, the face-to-face interviews were conducted at their houses. This out-reach approach helped the connection with those who seldom make their voice heard in public. The third phase begins in 2012. A community association in a fishing village was organized to strengthen the long-term relationship between the locals and the project team. Through more comprehensive community contacts and activities, it is believed that the locals' adaptive capacity can be built up and the future directions of community development can be pursued. Four types of the community activities were held during these years and there were interviews, group discussions, workshops and conferences. This bottom-up empowerment project not only provides various communication channels but also raises the awareness of climate changes of the locals. More actions will be taken for reaching the seven consensus goals during the third phase of the project. To sum up, this paper will explore the bottom-up, public participation approach for confronting the impacts of the global climate change.

The third chapter focuses on the modeling and simulation of mixing and sediment processes induced by tsunamis propagating upriver. The 26 December 2004 and 10 March 2011 tsunami disasters highlighted the rapid advance of the tsunami waters following rivers and canals causing massive damage deep inland. The leading edge of the tsunami formed a series of waves propagating upriver. This phenomenon is called 'shio-tsunami' (bore), 'kaisho' (tsunami-induced bore) or simply 'tsunami' (tidal wave) in Japan. After breaking, a tsunami wave propagating in shallow-waters is led by a bore. In these rivers and shallow-water bays, the propagation of the bores is associated with strong mixing and massive sedimentary processes upriver. This contribution reviews a number of tsunami-induced bore disasters before studying the tsunami propagation upriver. The aim of the contribution is to provide compelling evidences on the turbulent mixing and sediment processes induced by tsunami bores based upon some recent field observations as well as physical and numerical studies of tidal bores. It is shown that the propagation of the bore front modifies drastically the water column properties and the tsunami waters behind the front may carry a large amount of sediments, explaining the sedimentological and morphological changes observed after some recent disasters.

The fourth chapter leads us to the perspective of risk quantification for commercial nuclear power plants. Infrequently, it seems that a significant accident precursor or, worse, an actual accident, involving a commercial nuclear power reactor occurs to remind us of the need to reexamine the safety of this important electrical power technology from a risk perspective. Twenty-five years since the major core damage accident at Chernobyl in the Ukraine, the Fukushima reactor complex in Japan experienced multiple core damages as a result of an earthquake-induced tsunami beyond either the earthquake or tsunami design basis for the site. While the tsunami itself killed tens of thousands of people and left the area devastated and virtually uninhabitable, much concern still arose from the potential radioactive releases from the damaged reactors, even though there was little population left in the area to be affected. As a lifelong probabilistic safety analyst in nuclear engineering, even the chapter writer must admit to a recurrence of the doubt regarding nuclear power safety following Fukushima that he had experienced following Three Mile Island and Chernobyl. This article is his attempt to "recover" my personal perspective on acceptable risk by examining both the domestic and worldwide history of commercial nuclear power plant accidents and attempting to quantify the risk in terms of the frequency of core damage that one might glean from a review of operational history.

The fifth chapter discusses the future role of mathematical modeling for local energy shortage recovery after tsunamis. The natural catastrophe such as tsunamis, earth quake and hurricane will always bring a blow to local economy. The short-time energy shortage is one of the most serious damages in the disaster because it may cause local areas to suffer from delayed rescues, widespread power outages, tremendous economic losses, and even public safety threats. In such situation, agile energy dispatching through an effective energy transportation network, targeting the minimum energy recovery time, should be a top priority. Therefore, the application of computer aid design will be very essential to handle the above global concern. The mathematical modeling of economic optimization and dispatch scheduling will provide sufficient technical support to decision-makers. The computer simulation and optimization should provide optimal output solution with the huge input information of variable demand, local energy resources, energy facility capacity, economic sustainability and emergency under extreme conditions. Available renewable technology has provided more potential sources for local energy recovery while it has added the difficulty in the mathematical modeling of design and dispatch of multiple energy types in a wide regional area. In this chapter, the latest progress in this research area will be introduced and reviewed. Furthermore, the writer's newly developed mathematical models will also be introduced. It is developed for energy network optimization for the consideration of both energy recovery capability and economic factors. The efficacy of the developed methodology is demonstrated via case studies with in-depth discussions. Moreover, it is better to investigate a studied energy dispatch problem under uncertainties, such as energy loss percentage and energy supply to the system. Thus, the optimal dispatch time can be considered as a function of these uncertainties. The sensitivity of the minimum dispatch time with respect to uncertainty parameters is characterized by partitioning the entire space of uncertainty parameters into multiple subspaces. They provide decision makers a reference manual for the easy lookup of the possible impact from uncertainties and thus help conduct right decisions. Based on such information, the dynamic impact of different energy dispatch strategies can be evaluated, which will help decision makings under uncertainties. The developed methodology is the first prototype in the field of energy dispatch optimization under emergencies. Although it is based on some simplifications and assumptions, it has built a solid foundation for future in-depth study and real applications.

These chapters demonstrate the complexities involved in attempting to understand the impact of tsunamis to us in different aspects. Furthermore, it will also help us to improve our disaster management system in order to get a good preparedness for the future challenges. We hope that by the end of the book the readers will be able to thoroughly understand our above intentions. Actually it will become the first book in our planned book series for the topic of economic impact, disaster management and future challenges for the natural disasters and global climate change. Besides this book, our subsequent writing will focus on the engineering thinking on the other natural disasters such as avalanches, earthquake, volcanic eruptions, floods, lake overturn, blizzards, cyclonic storms, droughts, hailstorms, heat waves, tornadoes, hurricanes, wildfire, and global warming.

This book is the outcome of the research work carried out in the multiple universities and research institutes. Many colleagues and research associates have been involved in the respective research program. While a number of them are involved in this book writing as co-authors, we would like to take the opportunity to thank all those whose contribution and work has made this a reality. In particular, we would like to acknowledge the contributions of Gerald J. Kost, Corbin M. Curtis, William J. Ferguson, Richard F. Louie, Pratheep Katip, Sinite Yu, Kate Teong-Tsyr Wang, Hong-Yuan Lee, Bi-Liang Lin, Wen-Yao Hsu, Hubert Chanson, Pierre Lubin, and Raymond HV Gallucci. The appreciation should also be my parents Dan'an Cai and Mingyuan Shi for their always support for my research interest and their previous painstaking effort to bring me up.

Finally, we would like to thank NOVA Publishers and in particular Nadya S. Gotsiridze-Columbus, Stella Rosa, Jennifer Ramirez and Carra Feagaiga, for their enthusiastic support of this endeavor.

USA, May 2013

Tianxing Cai

INTRODUCTION

For most of us, the impression for the word "tsunamis" is only a combination of multiple elements of action movie, thriller, drama and science fiction, which can only be experienced through the movies such as *The Poseidon Adventure, The Impossible, Bait 3D, Haewundae, 2012*. However, we were indeed witnessing a tsunami in 2011. People are now very concerned and some are afraid of the realization of movie scenes in our real life. More fields have started to give more attention and thought on the tsunami itself and its impact on the economics, disaster management and future challenges. Here comes this book in order to disclose the close relationship between the tsunami and all the aspects of our real life: communities, markets, economic systems, industrial operation, natural environment, medical care, emergency response, and the whole societies. This book is timely and in some ways timeless; the issues discussed within its pages are matters that are of interest to all people across the world and really across time. The presented chapters focus on the theoretical positions and cognition about engineering thinking and suggestions for the tsunami in different segments of research fields: the innovative integration of geographic information system with point-of-care testing for emergency response and disaster management after tsunami (Chapter 1), the adaptive capacity building through public participation (Chapter 2), the modeling and simulation of mixing and sediment processes induced by tsunamis propagating upriver (Chapter 3), the perspective of risk quantification for commercial nuclear power plants (Chapter 4), and the future role of mathematical modeling for local energy shortage recovery after tsunamis (Chapter 5). These chapters demonstrate the complexities involved in attempting to understand the impact of tsunamis to us in different aspects.

Furthermore, it will also help us to improve our disaster management system in order to get a good preparedness for the future challenges.

In: Tsunamis
Editor: Tianxing Cai
ISBN: 978-1-62808-682-9

Chapter 1

EMERGENCY AND DISASTER POINT-OF-CARE TESTING IN SOUTHERN THAILAND AND PHANG NGA PROVINCE: TSUNAMI IMPACT, NEEDS ASSESSMENT, GEOSPATIAL PREPAREDNESS, AND FUTURE RESILIENCE

Gerald J. Kost,[1,2] Corbin M. Curtis[1], William J. Ferguson[1], Richard F. Louie[1] and Pratheep Katip[*3]

[1]Point-of-Care Center for Teaching and Research (POCT•CTR), Pathology and Laboratory Medicine, School of Medicine University of California, Davis, US

[2]Affiliate Faculty,

[3]College of Population Studies, Chulalongkorn University, Bangkok, Thailand

[*] Corresponding Author address: Gerald J. Kost, MD, PhD, MS, FACB Point-of-Care Testing Center for Teaching and Research (POCT•CTR), Pathology and Laboratory Medicine, 3455 Tupper Hall, School of Medicine, University of California, Davis, CA 95616, Phone: 530-574-3945; Fax 530-752-4548; Email gjkost@ucdavis.edu.

ABSTRACT

The 2004 Earthquake and Tsunami in Southeast Asia had devastating effects on coastal areas in Southern Thailand, where several thousand people lost their lives from drowning and associated injuries, while many more were displaced from their homes and businesses by destructive flooding, which disrupted roads and shut down transportation acutely. One of the hardest hit provinces was Phang Nga facing west to the Andaman Sea. After the acute Tsunami event, we conducted a damage assessment field survey of 4 provinces—Krabi, Phang Nga, Phuket, and Trang. Later we surveyed Phang Nga Province in detail to assess how healthcare facilities had recovered and adapted. Recently, we performed a needs assessment of diagnostic testing, including point-of-care (POC) testing, which is defined as testing at or near the site of care and includes both self-testing and noninvasive monitoring. We found significantly enhanced implementation of POC testing including increased use of noninvasive oxygen saturation monitoring in lieu of invasive in vitro blood gas testing. Currently, we are innovating terrain mapping and geographic information systems (GISs), which can be used to determine health centers, primary care units, community hospitals, and regional hospitals at risk for future tsunamis or other natural disasters. The GIS comprises a comprehensive database that includes all healthcare facilities in the province and also locates POC resources, backup, and preparedness factors, such as the critical distances from the most vulnerable population clusters along the coast and on islands to their sites of immediate, referral, and follow-up medical care. Elevation data in the GIS can be used to position heliports for future acute rescue and to determine critical land transportation paths for rescue. Overall, the small-world network of healthcare delivery in Phang Nga Province recovered nicely, but future leadership should adopt more advanced tools, such as GISs, to improve future preparedness and resilience.

INTRODUCTION

In several Southeast Asian nations, approximately 275,000-290,000 persons were dead or missing [1,2] as a result of the devastating Tsunami caused by a megathrust quake assessed at Richter 9.15-9.20 that struck Sunday morning, December 26, 2004 [3,4]. Another Sumatra-Andaman earthquake and Tsunami may occur in response to co-seismic stress, warranting improved preparedness and critical care responsiveness, including advancing the use of POC testing in Thailand and other countries [5]. In early 2005, we conducted field surveys [5,6] in areas severely damaged by the Tsunami in Southern Thailand (Figure 1) to assess the immediate disaster impact and response [7]. We returned in 2007 for direct inspection, followed by research through 2012 [7,8] to identify needs-based changes and to document new practice patterns for POC testing, particularly the use of oxygen (O_2) saturation monitors (pulse oximeters), which became widespread in coastal Phang Nga hospitals following the Tsunami.

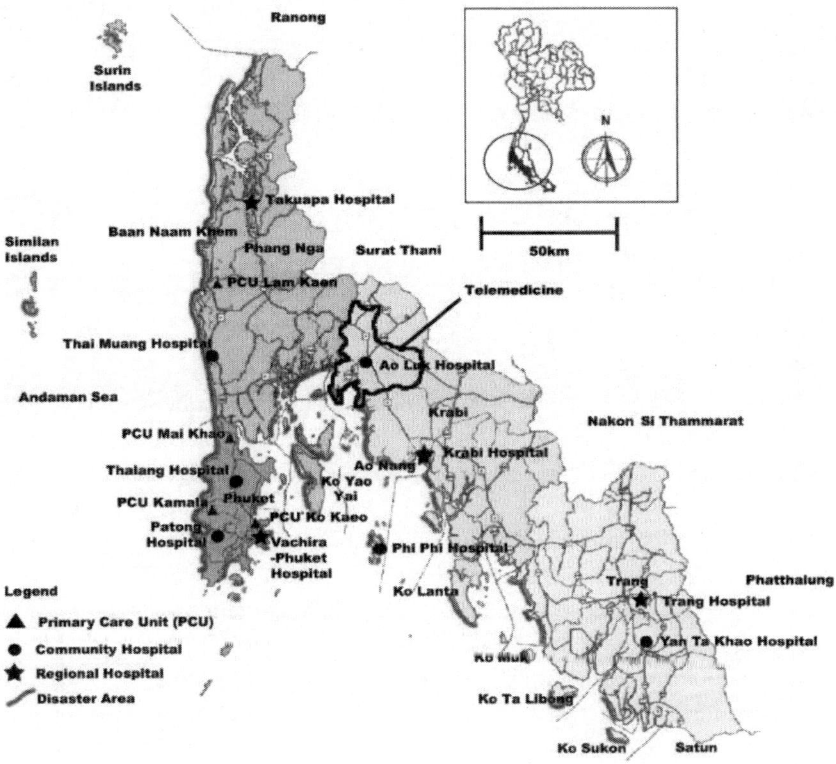

Figure 1. Tsunami Disaster Areas and Medical Sites in Southern Thailand. The Tsunami, up to ten meters in height, displaced a trillion tons of water, hit coastal and island areas hardest (red highlights), inundated transportation routes, sent victims scurrying to hospitals accessible nearby, and caused massive destruction of beach villages and resorts. Sea Gypsies living near and on the beach in the Surin Islands (to the Northwest) were saved by knowledge transmitted verbally from generation to generation of elders, who sounded life-saving warnings of impending danger just before the Tsunami hit shore. Ao Luk Community Hospital operated a unique telemedicine system (outlined) in Krabi province.

Most recently, we are developing geospatial tools that delineate areas at high risk from these "newdemics" [9], that is, unexpected and disruptive problems that affect the health of large numbers of individuals in a crowded world. These approaches can be used in conjunction with topographical positioning of POC testing in primary care (health care center) locations and community hospital hubs of high activity in low-resource settings to improve preparedness in healthcare small-world networks [10]. Coastal areas with high population density remain vulnerable, as evidenced by the complex emergency (i.e., earthquake and tsunami plus nuclear spill) that killed 18,000 Japanese in March 2011. Hence, the strategies illustrated in this chapter will enhance crisis standards of care [11] and local resiliency wherever newdemics might occur, whether caused by a tsunami or other natural and man-made phenomena.

Acute Response—Krabi, Phang Nga, Phuket, and Trang Provinces

Participants

Kamala Primary Care Unit (PCU) in Phuket Province was inundated by the Tsunami (Figure 2), closed at the time we visited, and remained inoperable for a prolonged period [6]. Patient care shifted to a rescue site in Kamala village 1 km inland. Medical problems included acute injuries, respiratory complications, infectious diseases, and psychological stress, which lingered for months following the Tsunami. Along the coast in Phang Nga Province (Figure 3) [7] located close to areas of heavy damage (Figures 4 and 5), Lam Kaen PCU was overwhelmed by victims with acute injuries, lacerations, drowning/ near drowning, and other medical crises resulting in nearly 100 deaths the same day. Primary Care Unit staff, POC equipment, drug supplies, and other resources were not adequate to deal with the intensity of the crisis. In four PCUs surveyed POC testing was limited early on to glucose meters (used for screening and monitoring of patients with diabetes), protein/ glucose urine dipsticks, spun hematocrit, urine pregnancy, and sometimes blood group typing, performed for youth identification card documentation.

Figure 2. Kamala Primary Care Unit (PCU). Built only slightly above the mean tide level, the Kamala PCU was inundated by the Tsunami (see the inset) and closed. The beach and Andaman Sea are visible a short distance away in the background (to the right).

Bedside testing consisted primarily of glucose meters used routinely by nurses in ERs and occasionally in ORs for immediate results during busy periods and at night. Pulse oximetry was used in ORs occasionally. Only one (17%) of six community hospitals surveyed in the four provinces provided blood gas/pH testing. Two lacked electrolyte testing. Another provided whole-blood analysis of Ca^{2+}, Na^{+}, K^{+}, and hemoglobin. Only one offered cardiac troponin (cTn) I testing, performed qualitatively in the ER. Physicians stated that all community hospitals should perform qualitative or quantitative cTn testing to facilitate prompt patient referral and/or treatment in lieu of overnight observation and delays in crisis triage. Infectious disease tests were limited. Three community hospitals lacked a microbiology laboratory. None had ICUs or CCUs, but nonetheless received numerous critically ill patients.

One community hospital close to the coast covered primary care for over 40,000 people and ten PCUs received up to 150 patient visits each weekday, but had only three physicians. The Tsunami hit mid-Sunday morning when only one physician, whom we interviewed, was on duty. Medicines depleted quickly, telecommunications failed, and waves disrupted roads forcing victims to go to the local hospital and delaying back-up help for five hours. The physician first determined whether victims were dead or alive and their state of consciousness, as large numbers arrived in automobiles, resort vans, and pickup trucks. Acute conditions included aspirations, intubations, trauma, pneumothorax, and lacerations.

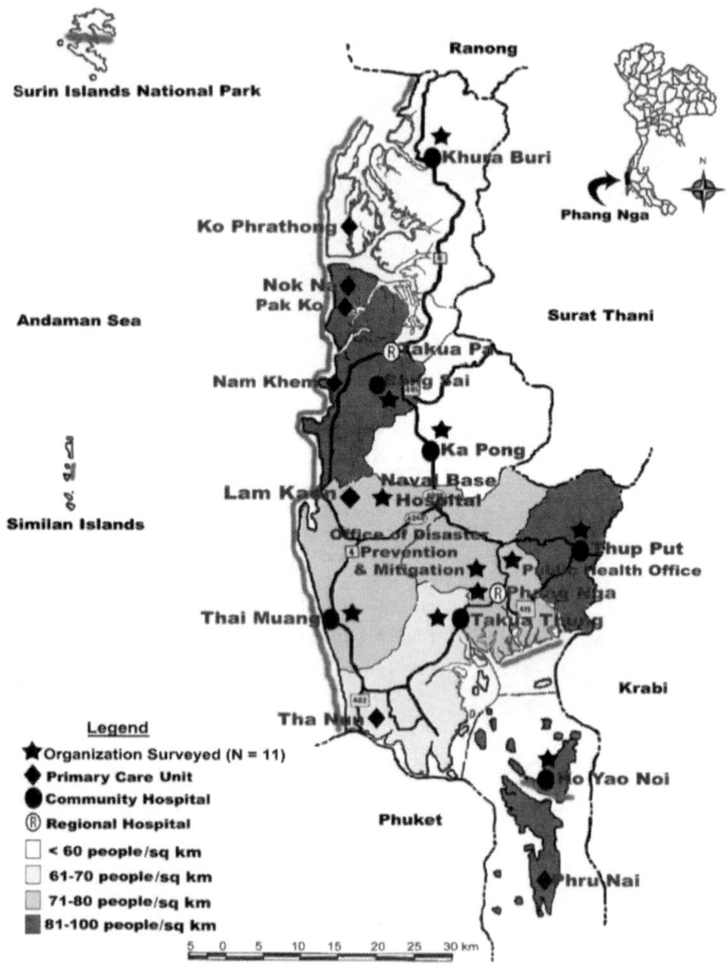

Figure 3. Phang Nga Province Small-World Network. This map shows population densities, geographic areas, small-world network main transportation routes, and locations of medical organizations affected directly or indirectly by the Tsunami. During a recent needs assessment survey, we contacted widely dispersed health care facilities (identified in the legend) including Community, Regional, and Naval Base Hospitals, the Office of Provincial Public Health, and the Office of Prevention and Mitigation.

Figure 4. High Impact Physical Damage Caused by the Tsunami. The Tsunami hit hardest along the heavily populated Khao Lak Region of the Phang Nga Coast between the Thai Muang and Takuapa Hospitals, sites that received thousands of injured victims. These photographs were taken at a well known five-star resort soon after the acute event. Local witnesses stated that the crest of the Tsunami exceeded 30 feet. The power of the Tsunami is evident in the catastrophic damage (top left), physical dislocation of whole buildings (top right), upending of heavily constructed villas (bottom left), and destruction of concrete and brick structures (bottom right), all with little or no warning. Hence, loss of human life was significant and motivated the installation in the next two years of warning signs (**Figure 5**) and local high volume audible alarms that alert tourists and Thais to seek high elevation when at risk from another Tsunami.

Figure 5. Typical Warning Sign Installed Near the Beach Following the Tsunami.

The one-room diagnostic laboratory in the community hospital lacked equipment for measurement of electrolytes, blood gases/pH, drug levels, and cardiac biomarkers. The physician decried the lack of electrolyte testing as a critical deficiency and stated that cTn testing was needed badly because patients with chest pain could not be triaged quickly or efficiently, resulting in excess mortality. A pharmacist rushed to the hospital to find analgesics and first aid supplies exhausted quickly. Lack of anesthesia equipment and specialists limited surgery, although the hospital had a small OR used routinely for C-sections, tubal ligations, and appendectomies. No helipad was available for airborne transport.

The Takuapa Regional Hospital [12] Emergency Room received 986 victims. With over twenty nationalities, lack of interpreters caused language difficulties and psychological stress for patients and staff. The extent of damage and severity of the crisis were underestimated. About 1 hr after the Tsunami hit, the ninc physicians who reported to the ER realized there were thousands injured. They activated the hospital's "major incident policy." All doctors were empowered to treat and refer patients. Frequent cases of aspiration were assessed in severity by POC pulse oximetry. The six-bed ICU quickly filled to twice capacity, then patients were sent elsewhere. Point-of-care testing was limited to glucose meters. No quantitative cardiac biomarker testing was available. The hospital had no helipad. Ambulances without POC testing were used for referral transport, not rescue. Deaths occurring during transfer may have been significant. Hospital staff could not accompany patients.

Ambulances met helicopters on a football field by city hall. Military helicopters and eventually, private jets, arrived at Phuket International Airport, although initially, the airport was closed. Interviewees stated back-up help arrived too slowly.

In the Phi Phi Islands (Krabi Province), which we surveyed, human losses were sudden and substantial—1,000 died with many bodies never found [6]. Phi Phi Hospital, a new community hospital that had been scheduled to open early in 2005 and was located on a beach isthmus between bays that focused the Tsunami (Figure 6), received the full brunt of the crest, which damaged the partially completed structure and removed tons of sand and stone walkways south of the foundation at water's edge. Diagnostic testing was limited to glucose meters, spun hematocrit, and urine dipsticks for protein and glucose. Staff consisted of one physician, two nurses, and one assistant. A pharmacist was killed while tending to patients when the tsunami arrived. The hospital boat was damaged.

Referral required one to two hours of land and sea transport to Vachiara-Phuket or Krabi Regional Hospitals on the mainland. No naval hospital ship arrived early to help. Interviewees stated airborne transport also did not meet acute needs.

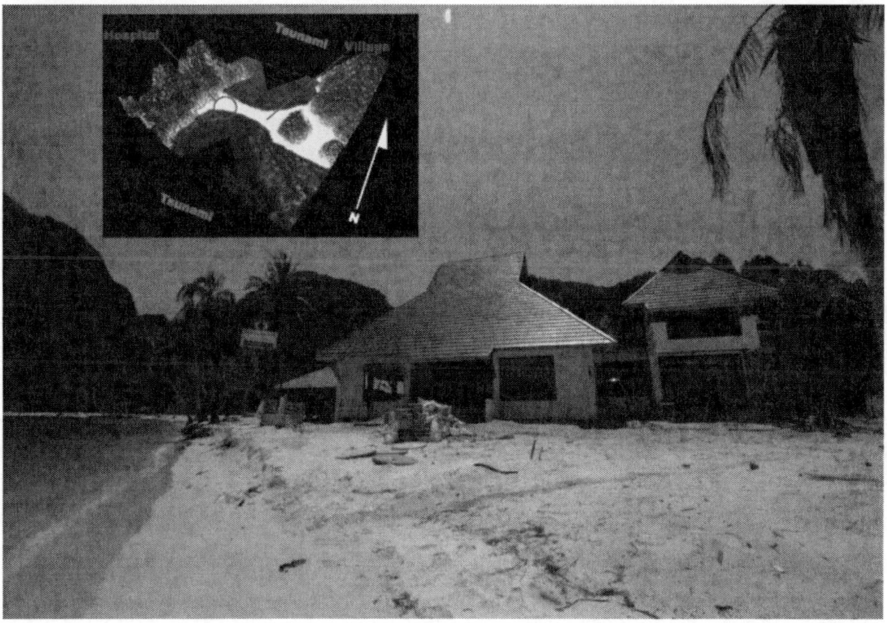

Figure 6. Phi Phi Island Hospital. The Tsunami swept north and south (see the inset), grew in size funneled by the two opposing bays, crossed the land bridge between the two mountainous masses, removed tons of sand in its path, and damaged the new hospital structure, which was left dangerously exposed and stranded on what remained of the beach. The building could serve well for primary care, day clinics, and tourist counseling, but local inhabitants felt that emergency and intensive care should be relocated to a safer site. Nonetheless, hospitals functions were installed. The top line of the bent sign in front of the hospital says in Thai, "'Roongpayaabaan Gaw (or Ko) Phi Phi," which means (bottom line), "Phi Phi (Island) Hospital." Before the Tsunami, the room with a sea view nearest the beach was intended to be an intensive care unit.

Using the Ao Luk telemedicine system (started October 2002), PCU nurses call into Ao Luk Community Hospital in Krabi Province by radio, demonstrate the patient's symptoms and signs via a computer monitor, and then receive actionable physician orders [13,14]. Outreach medical staff, including two dentists, rotate through amphur PCUs, deliver supplies, collect medial waste, and pick up blood samples (150 per week) for diagnostic testing (e.g., HIV, VDRL, cholesterol, triglycerides, creatinine, urea nitrogen, and Thalasemia screening) at the Ao Luk laboratory. Telemedicine allows, if necessary, referral of patients with serious chest pain directly to Krabi Regional Hospital 40 km south. Ao Luk staff want to implement POC cTn testing in the telemedicine-linked PCUs to improve the efficiency of triaging. Personnel were satisfied with the quality of care provided by distance medicine and felt that telemedicine-enabled small-world networks could improve care during future disasters.

It is interesting to note that in this devastating disaster, although their village located virtually at sea level in a beach area of the Surin Islands (an archipelago of 5 islands located in the Andaman Sea 55 km from the mainland and administratively part of Tambon Ko Phra Thong, Amphur Khuri Buri, in Phangh Nga Province—see Figures 1 and 3) was destroyed, no Mokens ("Sea Gypsies") died in the Tsunami [6].

Elders, whom we interviewed shortly after the acute event, remembered "the story of giant waves" that occurred each 250 years, observed bay waters receding, and then shouted warnings in time for all their village people to escape to higher elevation on surrounding hills. In contrast, there were no early warning systems elsewhere in the developed regions of Southern Thailand, although in the years that followed, escape route signs and audible warnings were installed along with advanced detection buoys at sea.

Lessons to be learned from the overall experience suggest that historically, resilience comes from self-reliance; socially, strategic planning will engender preparedness; and medically, a new "point of care culture" [15] should be explored and implemented in order to properly embed POC testing for on-site decision-making that can accelerate evidence-based diagnosis, triage, and treatment when conventional resources are not available, local community hospitals become isolated by flooding [16], and small-world network [17] infrastructure damage severs transportation and communication routes. Therefore, next we summarize POC resources during response and subsequent adaption that led to better preparedness, specifically in Phang Nga Province.

Table 1. POCT in Community and Regional Hospitals

Hospital	Community Hospital (CHs)							Regional Hospital (CHs)	
	Ko Yao Noi	Takua Thung	Thai Muang	Khura Buri	Thup Put	Ka Pong	Bang Sai	Takua Pa (TP)	Phang Nga (PN)
Beds	30	30	30	30	30	30	10	209	265
Medical doctors	1	4	4	3	3	2	1	19	30
Registered nurses	17	39	49	35	28	20	15	165	312
Lab staff / personnel	2	2	2	2	2	2	1	13	17
Primary Care Units	3	12	19	6	6	8	9	9	-
Intensive Care Unit	-	-	-	-	-	-	-	1	3
Neonatal Intensive Care Unit	-	-	-	-	-	-	-	1	1
Coronary Care Unit	-	-	-	-	-	-	-	-	-
• Cardiologist	-	-	-	-	-	-	-	-	-
Operating room	Yes	Yes	Yes	Yes	Yes	Yes	-	5	4
• Type of surgery	-	Minor surgery	Minor surgery	Minor surgery	Unavailable	Minor surgery	-	Both minor & major surgeries	Both minor & major surgeries
Labor room	Yes	Yes	Yes	Yes	Yes	Yes	-	Yes	Yes
Mother & Child Unit	-	Yes	-	-	Yes	Yes	-	1	4
Closest Hospital	Talang CH	Thai Muang CH	Takua Thung CH	Suksumran	Phang Nga RH	Bang Sai CH	Ka Pong CH	Bang Sai CH	Thup Put CH
• Distance (km)	60	12	15	18	20	25	20	12	22
• Travel time (hr/min)	1 hr, 30 mins	20 mins	20 mins	30 mins	30 mins	30 mins	30 mins	15 mins	20 mins
Referral Hospital	Vachira Phuket	Phang Nga RH	Takua Pa RH	Takua Pa RH	Phang Nga RH	Takua Pa RH	Takua Pa RH	Vachira Phuket	Vachira Phuket
• Distance (km)	50	40	65	56	20	32	7	140	140
• Travel time (hr/min)	1 hr, 20 mins	50 mins	1 hr	30 mins	30 mins	30 mins	20 mins	2 hrs	1 hr, 30 mins
Diagnostic Lab Services									
• Weekdays	08:30 - 16:30	08:30 - 16:30	08:30 - 16:30	08:30 - 16:30	08:30 - 16:30	07:30 - 16:30	08:30 - 16:30	08:30 - 16:30	08:30 - 16:30
• Weekends	Closed	08:30 - 16:30	08:30 – 12.00	08:30 - 16:30	08:30 - 16:30	08:30 - 12:00	08:30 - 16:30	08:30 - 16:30	On-call services
• On-call services	Yes	-	-	Yes	Yes	-	-	Yes	Yes
Microbiology Lab	-	-	-	-	-	-	-	Yes	Yes
POCT Status									
• Clinical wards	Glucose, Hct	Glucose, Hct Pregnancy, cTnT	Glucose, Hct	Glucose, Hct cTnT, VCT	Glucose, Hct VCT, Urine sp.g.	Glucose, Hct VCT	Glucose, Hct	Glucose	Glucose
• Emergency room	Glucose, Hct Tourniquet test	Glucose, Hct cTnT	Glucose, Hct cTnT	Glucose, Hct	Glucose, Hct VCT	Glucose, Hct	Glucose	Glucose	Glucose, Hct Urine ketone
• Operating room	-	-	-	-	-	-	-	Glucose	Glucose, Hct
• Labor room	Refer to lab	Urine glucose, Hct, Pregnancy	Hct, Urine albumin Urine glucose	Urine glucose Hct	Urine glucose Hct	Hct, Urine albumin Urine glucose	Hct, Urine albumin Urine glucose	Glucose	Hct, Urine albumin Urine glucose
• Intensive Care Unit	-	-	-	-	-	-	-	Glucose, Hct	Glucose, Hct
• Coronary Care Unit	-	-	-	-	-	-	-	-	-
Ambulance	2	2	3	3	2	2	1	4	5
Helicopter	-	-	-	-	-	-	-	-	-

Abbreviations: Hct, Hematocrit; VCT, Venous Clotting Time

Table 2. Critical Care and Cardiac Biomarker Testing

Test / Hospital	Community Hospitals (CHs)							Regional Hospitals (RHs)	
	Ko Yao Noi	Takua Thung	Thai Muang	Khura Buri	Thup Put	Ka Pong	Bang Sai	Takua Pa (TP)	Phang Nga (PN)
Critical Care • Blood gas analysis • Electrolyte analysis i) Specimen	- - -	- Na^+, K^+, Cl^- Total CO_2 Serum	- Na^+, K^+, Cl^- Total CO_2 Serum	- Na^+, K^+, Cl^- Total CO_2 Plasma	- Na^+, K^+, Cl^- Total CO_2 Serum	- Na^+, K^+, Cl^- Total CO_2 Serum	- Na^+, K^+, Cl^- Total CO_2 Serum	pH, pCO_2, pO_2 Na^+, K^+, Cl^- Total CO_2, Ca^{2+} Serum	pH ,pCO_2, pO_2 Na^+, K^+, Cl^-, Total CO_2, Ca^{2+} serum & plasma
Infectious Diseases Testing • Microbiology Lab • Bloodstream Infection • Malarial Infection	- - Both blood thin & thick films, CBC	- Yes Both blood thin & thick films, CBC	- - Both blood thin & thick films CBC	- Yes Both blood thin & thick films, CBC	- Refer to PN Both blood thin & thick films CBC	- - Both blood thin & thick films CBC, Dengue Ag Test Kit	- - Both blood thin & thick films CBC Tourniquet test Urine sp.g.	Yes Yes Both blood thin & thick films, CBC IgM/IgG Antibody Test Kit	Yes Yes Both blood thin & thick films, CBC IgM/IgG Antibody Test Kit
Hematology • CBC & Platelet Count • Hemostasis (Coagulogram) • Special Test	Yes Clotting Time OFT, DCIP	Yes Clotting Time ESR	Yes Clotting Time ESR, OFT,DCIP	Yes - -	Yes Refer to PN ESR, OFT, DCIP	Yes Clotting Time, TT, Bleeding Time ESR, OFT,DCIP Inclusion body	Yes - -	Yes Clotting Time, PT aPTT, Bleeding Time ESR, OFT, DCIP, LE cell, Hb typing	Yes Clotting Time, PT aPTT, Bleeding Time G-6-PD, ESR, OFT DCIP, LE cell, Hb typing, Reticulocyte count
CSF examination	-	Cell count /Cell Diff/Protein	Cell count	-	Refer to PN	-	-	Cell count / Cell Diff / Protein/Glucose/Gram stain/India ink	Cell count / Cell Diff / Protein/Glucose/
Blood Group Typing	ABO Group Rh	ABO Group Rh	ABO Group Rh	ABO Group Rh	ABO Group Rh	ABO Group Rh	ABO Group Rh	ABO Group Rh	ABO Group Rh
Blood Bank Section	-	-	-	-	-	-	-	- Cross matching - Direct / Indirect Coomb's test - Ab screening - Ab identification - Autologous blood transfusion **Transmissible-infectious diseases:** (Refer to Phuket Regional Blood Center) a) HIVAg b) HIVAb c) HBsAg d) Anti-HCV e) Syphilis	- Cross matching - Direct / Indirect Coomb's test - Ab screening - Ab identification **Transmissible-infectious diseases:** (Refer to Phuket Regional Blood Center) a) HIVAg b) HIVAb c) HBsAg d) Anti-HCV e) Syphilis

Test / Hospital	Community Hospitals (CHs)							Regional Hospitals (RHs)	
	Ko Yao Noi	Takua Thung	Thai Muang	Khura Buri	Thup Put	Ka Pong	Bang Sai	Takua Pa (TP)	Phang Nga (PN)
Cardiac Biomarkers Testing									
Qualitative cTnT (≥ 0.1 ng / mL)	-	Yes (ER)	Yes (ER)	Yes	Yes	-		-	Yes
Semi-Quantitative cTnT (0.1 - 2.0 ng/ mL)	-	-	-	-	-	-		cTnT h 232 analyzer	cTnT Reader analyzer
Qualitative cTn I	-	-	-	-	-	Yes		-	-
CK	-	-	-	Yes	-	-		Yes	Yes
CK-MB	-	-	-	-	-	Yes		Yes	Yes
LD (LDH)	-	-	-	-	-	-		Yes	Yes
Myoglobin	-	-	-	-	-	Yes		-	-
ProBNP	-	-	-	-	-	-		-	-
Needs/Improvements/ Developments	Needs microscope	Needs microscope	Not mentioned	Needs Automated Chemistry Analyzer	No needs or Improvements	1) Blood gas analyzer 2) Automated analyzer for thyroid function tests	No needs or Improvements	1) Hematocrit centrifuger 2) Microscope 3) Bilirubin tester 4) Electronic balance 5) pH meter 6) Refactometer 7) Blood cell differential counter 8) Autoclave	1) Automated Chemistry Analyzer 2) Automated analyzer for thyroid function tests

Abbreviations: CBC, Complete Blood Count; OFT, Osmotic Fragility Test; DCIP, Dichlorophenol-indophenol Precipitation Test for Hb E trait; ESR, Erythrocyte Sedimentation Rate; TT, Thrombin Time; PT, Prothrombin Time; aPTT, activated Partial Thromboplastin Time; INR, International Normalized Ratio

Recovery and Adaptation

Community and Regional Hospital Response. During the period of recovery from 2005 to 2013, we returned to Phang Nga Province and surveyed [7] basic physical resources, health care delivery, and diagnostics in nine hospitals (Table 1). A new laboratory was built in Thai Muang Community Hospital, and new equipment was available in Takua Pa Regional Hospital. All nine hospitals enhanced the use of POC O2 saturation monitoring. Table 1 also summarizes available resources in seven community hospitals and the two regional hospitals, including the number of beds, medical staff, laboratory services, intensive care units (ICUs), and operating rooms (ORs), as well as types of surgical procedures performed and referral distances and ambulance travel time required to reach neighboring hospitals. Critical values lists appeared inconsistently and were not complete [7].

Small-World Network Analysis (see Figure 3). The distance [mean (SD), range, median, and maximum] from 53 PCUs to seven community hospitals was 19 (12), 3-60, 15, and 60 kms, and from seven community hospitals to two regional hospitals, 39 (20), 7-65, 40, and 65 kms, respectively, while the time needed for ambulances to travel from PCUs to community hospitals was 25 (22), 4-120, 20, and 120 mins, and from community hospitals to regional hospitals, 43 (21), 20-80, 30, and 80 mins. From Ko Yao Noi Islands, the closest community hospital (Talang) and regional hospital (Vachira Phuket) are 60 and 50 kms, with ocean crossing times of approximately 1 hr 30 mins and 1 hr 20 mins, respectively. Overall, the sum of mean distances from PCU to community hospital to regional hospital is 58 (SD 24) kms, requiring a total mean travel time of 68 (SD 30) mins, but the respective maxima are 125 kms and 200 mins, and prior study soon after the Tsunami showed many sites were completely isolated without prospect of rapid backup [5,6].

Critical and Cardiac Care. Community hospitals did not have ICUs, while Takua Pa Regional Hospital had one, and Phang Nga Regional Hospital had three, plus a neonatal ICU. Hospitals networked with multiple PCUs. No hospitals had coronary care units (CCUs). CCU deficiencies paralleled the lack of cardiac biomarker tests available in community hospitals. No cardiac biomarker testing was available at Ko Yao Noi Community Hospital (Table 2) [7]. Qualitative ($\geq$0.1 ng/mL) cardiac troponin T (cTnT) testing was present at community and regional hospitals, except Ko Yao Noi and Ka Pong Community Hospitals, and also Takua Pa Regional Hospital. Qualitative cardiac troponin I (cTnI) testing was performed at Ka Pong Community

Hospital. Semi-quantitative cTnT testing was available at regional hospitals. Limited other cardiac biomarker tests were available in regional hospitals (e.g., CK, CK-MB, and LD) and rarely in community hospitals.

O_2 Saturation Monitoring. According to interviewees, after the Tsunami pulse oximeter use increased in all hospitals. The number of O_2 saturation monitors available (41 total) in the seven community hospitals (Figure 7) surveyed ranged from 3 to 12 of different brands and formats with mean 5.9 (SD 3.1) and 83% portable, 17% built-in [7]. Bang Sai, Takua Thung, and Thai Muang were the only three community hospitals that used O_2 saturation monitors in birthing areas. All community hospitals have access to one or more in the ER and/or ambulances. Phang Nga community hospitals on average have fewer ($P < 0.05$) O_2 saturation monitors compared to community hospitals in the control province in northern Thailand, Chiang Rai, where the mean was 11.3 (SD 4.8) with 56% portable, 44 % built-in.

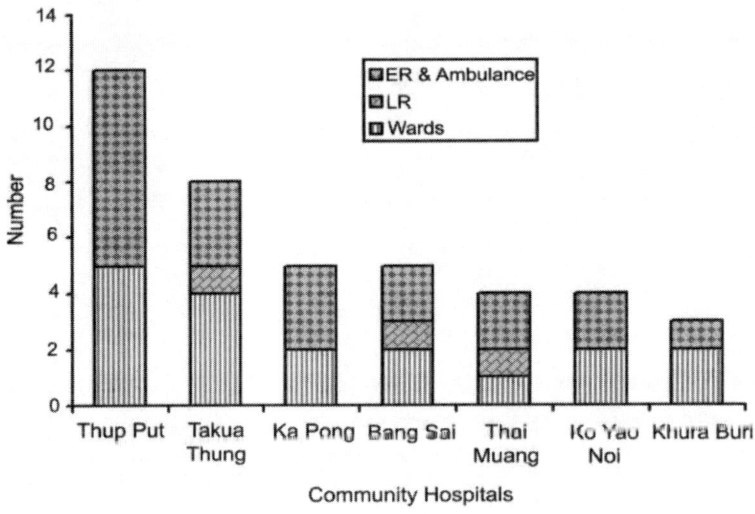

Figure 7. Distribution of O_2 Saturation Monitors in Community Hospitals. This bar graph shows the number of O_2 saturation monitors available in emergency rooms (ER) and ambulances, labor rooms (LR), and general wards in community hospitals in Phang Nga Province. Average utilization was not as great as it was in a control comparison province ($P < 0.05$), Chiang Rai, in northern Thailand.

The two regional hospitals in Phang Nga used O_2 saturation monitors frequently (Figure 8) [7]. The total number available is 105. Phang Nga Regional Hospital had 44 (52% built-in, 48% portable), while Takua Pa

Regional Hospital had 61 (56% built-in, 44% portable). In both regional hospitals, the fewest O_2 saturation monitors were available for obstetrics (i.e., labor rooms), but most equipment was portable and generally used anywhere needed. We found no continuous hemoglobin monitoring devices in use in community of regional hospitals, despite occurrence of Dengue hemorrhagic fever local epidemics that called for rapid packed red blood cell transfusions in critically ill patients. Figure 9 shows one pulse oximeter model observed in Phang Nga (top left) after the Tsunami and the kaleidoscope of other monitoring technologies available for future use [7].

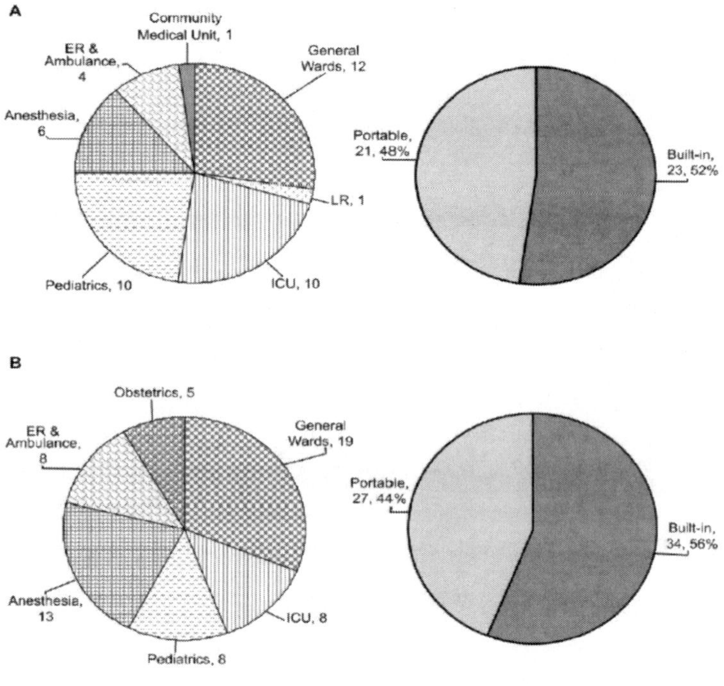

Figure 8. O_2 Saturation Monitoring at Phang Nga (A) and Takua Pa (B) Regional Hospitals. The pie charts illustrate the distribution of O_2 saturation monitors available in different hospital areas. Several devices were donated following the Tsunami. Takua Pa is closer to the coast (see small-world network, **Figure 3**) than Phang Nga and therefore, will benefit from more devices. The pie charts to the right show percentages built into permanent monitoring stations. Portable, handheld, and fingertip pulse oximeters are used during ambulance transport. The control RH in Chiang Rai had more of these POC devices.

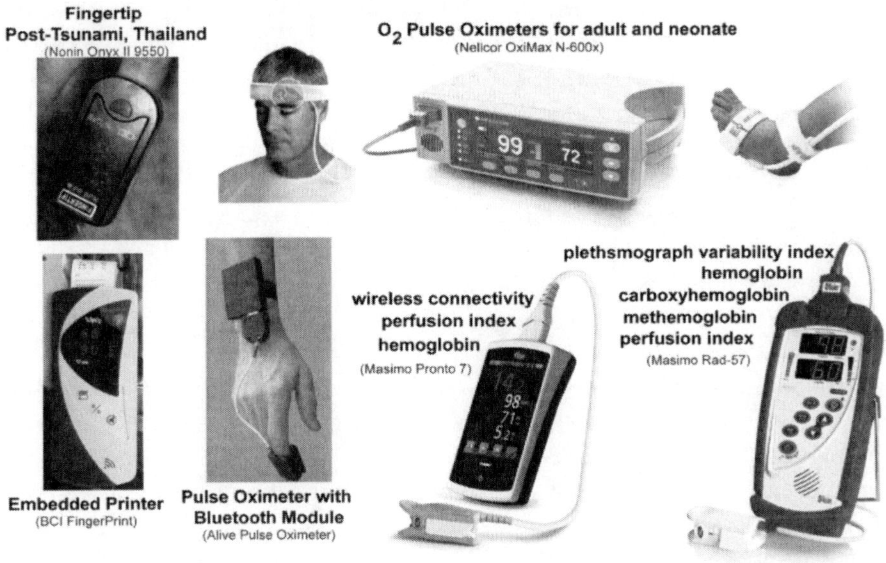

Figure 9. Current Approaches to O_2 Saturation and Hemoglobin Monitoring. Pulse oximeters measure O_2 saturation and heart rate noninvasively and incorporate useful features that this collage illustrates, as follows, clockwise from the top left: a) Nonin Onyx II 9550 in a community hospital after the Tsunami (photo by GK)—model 9560 has Bluetooth wireless connectivity (www.nonin.com); b) Nellcor OxiMax N-600x—works for adults and neonates (www.nellcor.com); c) Masimo Rad-57—utilizes Rainbow SET Pulse CO-Oximetry to continuously measure hemoglobin, carboxyhemoglobin, methemoglobin, O_2 saturation, plethsmograph variability index, and perfusion index (www.masimo.com); d) Masimo Pronto 7—offers 802.11 b/g and Bluetooth wireless connectivity and measures hemoglobin and perfusion index; e) Alive Pulse Oximeter—connects to a Bluetooth transmitter for wireless connectivity (www.alivetec.com); and f) BCI FingerPrint—features an embedded printer (www.smiths-medical.com). [Notes: Perfusion index (PI) is the ratio of pulsatile blood flow to nonpulsatile blood and surrounding tissues, which absorb a constant amount of light (DC) transmitted by the pulse oximeter. Pulsatile blood, however, absorbs a variable amount of light (AC) corresponding to the changing volume of blood with each pulse. The perfusion index then is expressed as: PI = (AC/DC) x 100%, which indicates the level of peripheral perfusion at the site of the monitor. The higher the PI, the more reliably the pulse oximeter will measure oxygen saturation. Some Masimo devices (e.g., Rad-57) also determine the plethysmographic variability index (PVI), a measure of changes in PI over a time interval, defined as: PVI = [(PIMax–PIMin)/(PIMax)] × 100%. PVI dynamically trends changes in PI to quantify changes in physiological states].

Other Testing. Staff performed POC glucose and hematocrit (Hct) testing in clinical wards and the ER, OR, LR, and ICU in Phang Nga community and regional hospitals (Table 1) [7]. The seven community hospitals primarily used glucose testing in clinical wards, ER, and LR, except Ko Yao Noi Community Hospital, for which there is no labor room. Takua Thung and Thai Muang Community Hospitals performed a qualitative cTnT test in the ER (Table 1) [7]. Pregnancy testing was performed on clinical wards in Takua Thung Community Hospital. Khura Buri, Thup Put, and Ka Pong Community Hospitals performed venous clotting time testing in clinical wards. In addition, a tourniquet test for bleeding time was performed at Ko Yao Noi Community Hospital in the ER.

Blood Gases, pH, and Electrolytes. No blood gas testing was offered in any of the seven community hospitals (Table 2) [7]. Blood gas (pCO_2, pO_2) and pH analyses were provided in regional hospitals where ORs were present and general anesthesia was performed. Electrolytes were measured in laboratories in all surveyed hospitals, except Ko Yao Noi CH. Takua Thung, Thai Muang, Thup Put, Ka Pong, Bang Sai Community Hospitals, and Takua Pa Regional Hospital measured electrolytes (Na^+, K^+, Cl^-, Ca^{2+}, and/or Total CO_2) in serum, while Khura Buri Community Hospital measured these analytes in plasma and Phang Nga Regional Hospital, in both serum and plasma. Just after the Tsunami [5,6] only 50% of the community hospitals provided Na^+/Cl^-/Total CO_2 measurements, and 67%, K^+. Electrolytes expanded as a result of a government purchasing program for the whole province, a key improvement in the standard of care and small-world network preparedness.

Hematology and Coagulation. Hematology tests included complete blood count (CBC) and platelet count in all nine hospitals (Table 2), also an improvement since the Tsunami [7]. Both Takua Pa and Phang Nga Regional Hospitals additionally performed hemoglobin (Hb) typing. Clotting time measurement for hemostasis was found in half of the community hospitals. Hemostasis assays in regional hospitals included clotting time, bleeding time, prothrombin time (PT), and activated partial thromboplastin time (aPTT).

There also were special tests [e.g., osmotic fragility (OF) and dichlorophenol-indophenol precipitation test (DCIP)] at some hospitals, except Khura Buri Community Hospital. ABO and Rh blood typing were available in all hospitals. The community hospitals lacked blood banks (BB) and half of them also do not offer cerebral spinal fluid (CSF) examination. Some laboratories in community and regional hospitals need equipment, devices, and automated analyzers for microscopy, automated chemistry, blood gases, and thyroid testing, plus support (e.g., electronic balance).

Microbiology and Infectious Diseases. Table 3 summarizes microbiology Infectious disease tests (e.g., malaria by microscopy and Dengue IgG/IgM testing) [7]. Khura Buri and Thup Put Community Hospitals, as well as Takua Pa and Phang Nga Regional Hospitals employed an IgG/IgM antibody test kit for Dengue Fever detection. In community hospitals, Dengue Fever hemolysis was monitored indirectly by determining the complete blood count (CBC) and Hct. When necessary for fast therapeutic turnaround time (time from order to treatment), one community hospital moved a small centrifuge (spun hematocrit) to the ward so that spun hematocrits could be performed hourly or as necessary. Ka Pong Community Hospital performed Dengue Ag testing, and Bang Sai Community Hospital performed tourniquet testing for the appearance of petechiae, in order to aid in diagnosing Dengue hemorrhagic fever outbreaks.

Primary Care Units. There are 53 PCUs in Phang Nga Province. Table 4 presents demographics, medical staff/personnel, service schedules, location, POC testing status, care plan notes, and survey results for seven PCUs surveyed [7]. All PCUs visited are situated in close proximity to the ocean, making them high risk for tsunamis and typhoons. In fact, Pru Nai, Nok Na, Pak Ko, and Ko Prathong PCUs are all located on islands. The population density is disproportionate to the staff resources available. For example, Pru Nai, the oldest island PCU listed, had the largest patient volume (60 per day), but has no physician on staff. Two PCUs, Nok Na and Ko Prathong, had no registered nurses available.

Point-of-care glucose testing remained ubiquitous among the PCUs, where we found each monitors glucose in diabetic patients differently. Some provide annual care plan notes while others, monthly. Specific tests for drugs of abuse (e.g., methamphetamines, morphine, and marijuana) were available at the Lam Kaen, Pak Ko, and Ko Prathong PCUs. Most of the PCUs surveyed perform detection of pathogens (bacteria) in both water and food samples by referring those to reference laboratories using conventional culture methods, including biochemical tests.

Table 3. Microbiology and Immunology Services

Test /Hospital	Community Hospitals (CHs)							Regional Hospitals (RHs)	
	Ko Yao Noi	Takua Thung	Thai Muang	Khura Buri	Thup Put	Ka Pong	Bang Sai	Takua Pa (TP)	Phang Nga (PN)
1.Microbial Testing									
Gram stain	Yes	Yes	Yes	Yes	Yes	Yes	Yes	Yes	Yes
Acid fast stain (AFB)	Yes	Yes	Yes	Yes	Yes	Yes	Yes	Yes	Yes
India ink preparation	-	-	-	-	-	-	-	Yes	Yes
KOH preparation	Yes	Yes	Yes	Yes	Yes	Yes	Yes	Yes	Yes
Wet smear	-	Yes	Yes	Yes	Yes	Yes	Yes	Yes	Yes
Fungal culture	-	-	-	R^1	R^1	R^1	R^1	Yes	-
Culture-aerobe	-	R^1	R	R^1	R^1	R^1	R^1	Yes	Yes
Cluture-anaerobe	-	R^1	R	R^1	R^1	R^1	R^1	R^1	R^2
Hemoculture	-	R^1	R	-	R^1	R^1	R^1	Yes	Yes
TB-culture	-	R^1	R	R^1	R^1	R^1	R^1	Yes	R^2
Urinanalysis	Yes	Yes	Yes	Yes	Yes	Yes	Yes	Yes	Yes
Fecal examination	Yes	Yes	Yes	Yes	Yes	Yes	Yes	Yes	Yes
2. Specimen									
Blood	Yes	Yes	Yes	Yes	Yes	Yes	Yes	Yes	Yes
CSF	-	R^1	Yes	-	R^1	Yes	-	Yes	Yes
Throat swab	R	R	R	-	R^1	R^1	-	Yes	Yes
Rectal swab	R	R^1	R	R^1	R^1	R^1	-	Yes	Yes
Stool	Yes	Yes	Yes	Yes	Yes	Yes	Yes	Yes	Yes
Sputum	Yes	Yes	Yes	Yes	R^1	Yes	-	Yes	Yes
Urine	Yes	Yes	Yes	Yes	Yes	Yes	Yes	Yes	Yes
Pus	-	R^1	R	Yes	R^1	R^1	Yes	Yes	Yes
Fluids	-	R^1	R	Yes	R^1	Yes	Yes	Yes	Yes
3. HIV Profile									
Serodia HIV	Yes	Yes	Yes	Yes	Yes	Yes	-	Yes	Yes
Determine HIV 1/2	Yes	Yes	Yes	Yes	Yes	Yes	Yes	Yes	Yes
SD Bioline HIV 1/2	-	Yes	-	Yes	Yes	Yes	-	Yes	Yes
$CD4^+$ count	R	-	-	R^3	R^1	R^2	R^2	R^5	Yes
Viral road	R	-	-	R^4	R^3	R^3, R^4	R^3	R^4	-
Anti-HIV	Yes	Yes	Yes	Yes	Yes	Yes	Yes	Yes	Yes

Test /Hospital	Community Hospitals (CHs)							Regional Hospitals (RHs)	
	Ko Yao Noi	Takua Thung	Thai Muang	Khura Buri	Thup Put	Ka Pong	Bang Sai	Takua Pa (TP)	Phang Nga (PN)
4. Immunological Tests									
i) Syphilis									
VDRL / RPR	Yes	Yes	Yes	Yes	-	Yes	Yes	Yes	Yes
TPHA	R	Yes	Yes	Yes	Yes	-	Yes	Yes	Yes
FTA-Abs	-	-	-	-	-	-	R^1	Yes	-
ii) Hepatitis Profile									
HBsAg	Yes	Yes	Yes	Yes	Yes	Yes	Yes	Yes	Yes
HbeAg	-	-	-	-	-	-	R^1	Yes	-
Anti-HBc	R	R^2	Yes	Yes	Yes	-	R^1	Yes	Yes
Anti-HBs	Yes	Yes	Yes	Yes	Yes	Yes	Yes	Yes	Yes
Anti-HBe	-	-	-	-	-	-	R^1	Yes	-
HepatitisA Ab (IgM)	-	-	-	-	-	-	-	-	-
HepatitisC Ab	R	-	-	-	-	R^1	R^1	Yes	-
iii) Influenza									
Type A	-	-	-	-	R^4	R^1	-	R^3	R^2
Type B	-	-	-	-	R^4	R^1	-	R^3	R^2
5. Other tests									
Leptospira	-	-	-	Yes	Yes	Yes	Yes	Yes	Yes
Widal Test	-	Yes	-	Yes	Yes	-	Yes	R^2	Yes
Weil Felix Test	-	Yes	-	Yes	Yes	-	Yes	Yes	Yes
Anti-DNA	R	R^2	-	R^2	R^2	-	R^1	Yes	R^1
Anti-Nuclear Factor	-	-	-	R^2	R^1	R^1	R^1	R^2	R^1
Meliodosis Titer	-	R^1	-	R^1	-	-	-	Yes	R^2
ASO Titer	-	R^1	-	R^1	R^1	-	R^1	Yes	Yes
Rhuematiod Factor	R	Yes	-	Yes	R^1	-	R^1	Yes	Yes
Dengue fever	CBC	CBC	CBC	CBC, IgM/IgG	CBC, IgM/IgG	CBC, DengueAg	CBC, Tourniquet	CBC, IgM/IgG	CBC, IgM/IgG
R = Refer	R: PN	R^1: PN R^2: Dhonburi Lab Center	R: TP	R^1: TP R^2: Dhonburi Lab Center R^3: PN R^4: Office of Disease Prevention & Control, 10th, Nakhonsithammarat	R^1: PN R^2: Dhonburi Lab Center R^3: Roche cop. R^4: Department of Medical Sciences	R^1: TP R^2: Department of Medical Sciences R^3: Vachira Phuket hospital R^4: Office of Disease Prevention & Control, 21st, Nakhonsithammarat	R^1: TP R^2: PN R^3: Office of Disease Prevention & Control, 10th, Nakhonsithammarat	R^1: Bureau of TB R^2: Dhonburi Lab Center R^3: Department of Medical Sciences R^4: Ramathibodi hospital R^5: PN	R^1: : Dhonburi Lab Center R^2: Department of Medical Sciences

Table 4. Tsunami Follow-up Role of POCT in Primary Care Units (PCUs)

Primary Care Units (PCUs)	Pru Nai	Tha Nun	Lam Kaen	Nam Khem	Nok Na	Pak Ko	Ko Prathong
Year established	1971	1983	1997	1995	1978	1997	2006 (afterTsunami)
Demographics							
• Population density (per km²)	108.78	160.78	94.7	91	91	60	23
• PCUs in district	3	12	9	8	9	9	6
• Telemedicine	-	-	Yes	-	-	-	-
• Pop. covered by site	5,525	6,431	7,000	3,018	409	568	902
• Patient volume / day	60	15-20	30	30	6	10-25	3
• Referral hospital	Vachira Phuket	Takua Thung CH	1) Thai Muang CH 2) Takua Pa RH	Takua Pa RH	Takua Pa RH	Takua Pa RH	Khura Buri CH
i) Distance to hospital (km)	30	6	1) 30-32 2) 35-40	7.8	25	13	15
ii) Travel time (hr/min)	1 hr	5-10 mins	1) 1 hr 2) 1 hr, 30 mins	38 mins	1 hr	25 mins	1 hr
Saff / Personnel							
• Medical doctor	-	-	-	-	-	-	-
• Registered nurse	1	3	3	2	-	1	-
• Technical nurse	-	-	-	-	1	-	-
• Medical science assistant	-	-	1	-	-	-	-
• Administrator	1	1	1	1	1	1	1
• Health staff	1	2	2	2	-	1	1
Service Schedule							
• Working Days (Mon – Fri.)	08:30 – 16:30	On-call services	08:30 – 16:30	08:30 – 16:30	08:30 – 16:30	08:30 – 16:30	On-call services
• Weekends	08:30 – 16:30	On-call services	08:30 – 16:30	On-call services	On-call services	On-call services	On-call services
• On-call services	Yes	Yes	Yes	Yes	Yes	Yes	Yes

Primary Care Units (PCUs)	Pru Hai	Tha Nun	Lam Kaen	Nam Khem	Nok Na	Pak Ko	Ko Prathong
Location	On Ko Yao Noi Island	~3 kms away from Andaman sea	~1 km away from Andaman sea	~1 km away from coastal area	On Ko Khor Khao Island	On Ko Khor Khao Island	On Ko Prathong Island
POCT Status							
• Glucose meter	7	6	6	1	1	1	2
• Spun hematocrit	1	1	1	1	1	-	1
• Urine protein testing	Yes	Yes	Yes	-	Yes	Yes	Yes
• Urine glucose testing	Yes	Yes	Yes	-	Yes	Yes	Yes
• Urine pregnancy testing	Yes	Yes	Yes	Yes	Yes	Yes	Yes
• ABO blood group testing	Yes	-	Yes	-	-	-	-
• Drugs of abuse testing							
◦ Methamphetamine	-	-	Yes	-	-	-	-
◦ Morphine	-	-	-	-	-	-	-
◦ Marijuana	-	-	Yes	-	-	Yes	Yes
◦ Amphetamine	-	-	-	-	-	Yes	Yes
• Detection of pathogen(s) in water and food samples	Yes	Yes	Yes	Yes	Yes	Yes	Yes
Care Plan Notes	◦ Monitoring of blood glucose in diabetic patients monthly.	◦ Monitoring of blood glucose in diabetic patients monthly. ◦ Screening for DM & HT in local population aged over 40. ◦ Providing health care delivery to the local population. a) vaccines for children aged 0-4 monthly. b) vaccines for pregnant women weekly.	◦ Monitoring of blood glucose in diabetic patients monthly.	◦ Screening for DM & HT aged over 35 yearly. ◦ Screening for breast cancer, cervical carcinoma, cardiovascular diseases, and stroke yearly.	◦ Family planning on Mondays. ◦ Screening for DM, at the age of over 35 years on Tuesdays. ◦ Provide necessary vaccines for children. ◦ Antenatal Care services	◦ Focus on delivery of health care for its locals. ◦ Focus on nurse aids and surveillance & disease control ◦ Screening for DM yearly.	**There are 4 Projects now** 1) Screening for DM & HT yearly. 2) Screening for cervical carcinoma yearly. 3) Dental care 3 times / year. 4) Controlling malarial epidemics yearly.

Abbreviations: DM, Diabetes Mellitus; HT, Hypertension

No POC cardiac biomarker, hemostasis, or HbA1c tests were available in any PCUs. Urinalysis, consisting of gross protein (but not microalbuminuria) and glucose testing, as well as pregnancy testing, was available in all PCUs, except Nam Khem, for which only urine pregnancy testing available. Spun hematocrit was performed in all PCUs, except Pak Ko. Several PCUs have implemented changes in an attempt to improve disaster preparedness, including additional support buildings and a new health care center (see Table 4) [7]. Additionally, Lam Kaen PCU installed telemedicine to the nearest community hospital, Thai Muang. In summary, province health care facilities improved resources broadly and appropriately, and also specifically at the POC, but sometimes unevenly. In view of the partial recovery, we surveyed the Province again in 2011-12 to establish local views of actual needs, in order to help improve future preparedness.

Needs Assessment—Phang Nga Province

Formal needs assessment determines when new POC technologies should be developed, which tests and test clusters must be available at the point of need, what designs are most useful to people who will perform the testing, how to deploy POC testing in the small-world network, and demographics for implementation [8]. Coastal Phang Nga Province lies in the seventh public health region of Thailand and has a population of 252,385 [18]. Figure 3 shows the population distribution by district (amphur) [7]. The number of people per hospital bed, medical doctor, registered nurse, technical nurse, pharmacist, and medical technologist inclusive of seven community and two regional hospitals is 411, 4,428, 460, 4,428, 7,648, and 2,696, respectively [19,20]. Phang Nga Province was devastated by the Tsunami that hit shore bluntly, destroyed inhabited structures (see Figure 4), and results in 4,186 deaths, 5,597 injuries, and 1,935 missing [21]. However, over the past decade the province has recovered substantially. Enhanced awareness of the need for preparedness generated enthusiasm among health professionals for our recent needs assessment survey, which we summarize next.

Small-World Network (see Figure 3). We determined locations of organizations surveyed (N = 11) and geographic relationships in the Phang Nga small-world network. The distance [mean (SD), range, median, and maximum] from 53 Primary Care Units (PCUs) to seven community hospitals was 19 (12), 3-60, 15, and 60 kms, and from seven community hospitals to two regional hospitals, 39 (20), 7-65, 40, and 65 kms, respectively, while the time needed for ambulances to travel from PCUs to community hospitals was 25 (22), 4-120, 20, and 120 mins, and from community to regional hospitals, 43 (21), 20-80, 30, and 80 [7,22,23].

Table 5. Pathogen Test Menus for Emergency and Disaster Care

Objective	Weighted Score	Pathogen
A. Civil disaster infections (n = 24)	53	*Staphylococcus aureus*
	42	*Klebsiella* species
	41	Dengue fever virus
	36	*Pseudomonas aeruginosa*
	34	HIV 1 and 2
	30	Hepatitis B
	28	*Enterobacter* species
	23	*Vibrio cholerae*
	23	*Plasmodium vivax*
	20	*Plasmodium falciparum*
	18	*Streptococcus pyogenes*
B. Respiratory pandemics (n = 24)	60	SARS
	60	Avian influenza (H5N1)
	45	Respiratory syncytial virus
	43	*Streptococcus pneumoniae*
	42	*Influenza A/B*
	41	*Mycobacterium tuberculosis*
	37	*Haemophilus influenza*
	35	*Mycoplasma pneumoniae*
	34	*Staphylococcus aureus*
	33	*Klebsiella* species
	26	Pandemic (H1N1) 2009 influenza
C. Bloodstream infections (n = 24)	95	*Streptococcus pneumoniae*
	85	*Staphylococcus aureus*
	75	*Escherichia coli*
	72	*Pseudomonas aeruginosa*
	65	*Streptococcus* species
	52	*Klebsiella* species
	48	Methicillin resistant *Staphylococcus aureus*
	43	*Enterobacter* species
	31	*Acinetobacter baumannii*
	30	Coagulase negative *Staphylococcus*
D. Emergency blood donor screening (n = 24)	155	Hepatitis B
	150	HIV 1 and 2
	138	Hepatitis C
	68	Epstein-Barr virus
	61	Dengue fever virus
	57	*Cytomegalovirus*
	49	Parvovirus B19
	49	Chikungunya virus
	43	Human T-cell lymphotropic virus 1 and 2 (HTLV 1 and 2)

Pathogen Detection. Needs assessment identified several pathogen test menus preferred by health professional respondents (N = 24) by ranking weighted scores in cases of potential civil disaster infections, respiratory pandemics, blood stream infections (sepsis), and emergency blood donor screening (Table 5) [8]. For the civil disaster setting, Staphylococcus aureus had the highest weighted score (53); for respiratory pandemics, SARS (60); in blood stream infections, Streptococcus pneumoniae (95); and for emergency blood donor screening, hepatitis B virus (155).

Test Clusters and Environmental Conditions. Figure 10 illustrates weighted scores of the top twenty tests that respondents selected for emergencies and disasters [8]. Complete blood count (CBC), electrolytes/chemistry, blood bank (blood group), oxygen saturation (by pulse oximeter), and hematocrit were selected as the top five POC tests to have in Phang Nga Province. Figure 11 illustrates eight environment conditions that health professional respondents considered important [8]. Temperature, vibration, humidity, and impact shock were the four most important ones when designing POC devices used during extreme conditions in local environments.

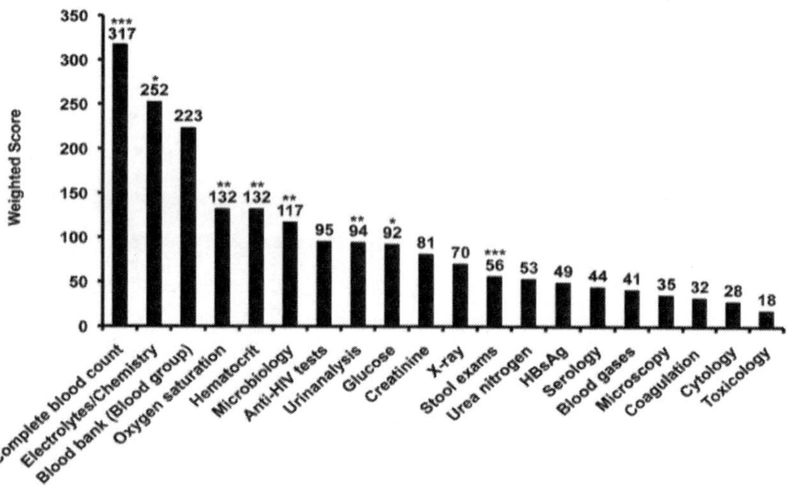

Figure 10. "Wish List" Priorities for Diagnostic Tests during Disasters. Health professionals in Phang Nga Province selected complete blood count, electrolytes/chemistry, blood bank (blood group), oxygen saturation (by pulse oximeter), and hematocrit as the 5 most important POC tests to have in the Phang Nga Province SWN. These tests differed significantly from 1 or more of the 3 other asterisked tests (urinalysis, glucose, and stool exams) shown in the bar graph. [*** P<0.001, ** P<0.01, * P<0.05].

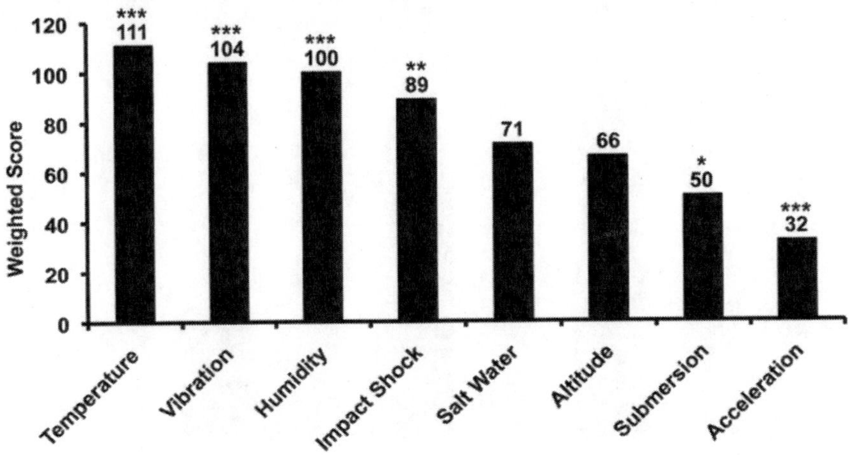

Figure 11. Environment Considerations for Future POC Devices during Extreme Environmental Conditions. Phang Nga health professionals chose temperature, vibration, humidity, and impact shock as the 4 most important environmental conditions to consider when designing POC devices for high risk settings, all 4 of which differed significantly from one or both of the other asterisked conditions (submersion and acceleration). [*** P<0.001, ** P<0.01, * P<0.05].

Table 6. Device Design: Comparison of Sample Collection Methods for Disaster Settings

N = 23		Criteria	N = 21	
Vacutainer® +Hub	Test Cassette		Coupled Sampling	Direct Sampling
16	7	Lowest risk of contamination	6	15
14	9	Lowest risk of operator error	10	11
7	16	Lowest risk from biohazard disposal	5	16
6	17	Lowest recurring cost	8	13
10	13	Overall, the best sample method	6	15
53	**62**	**Overall scores**	**35**	**70**

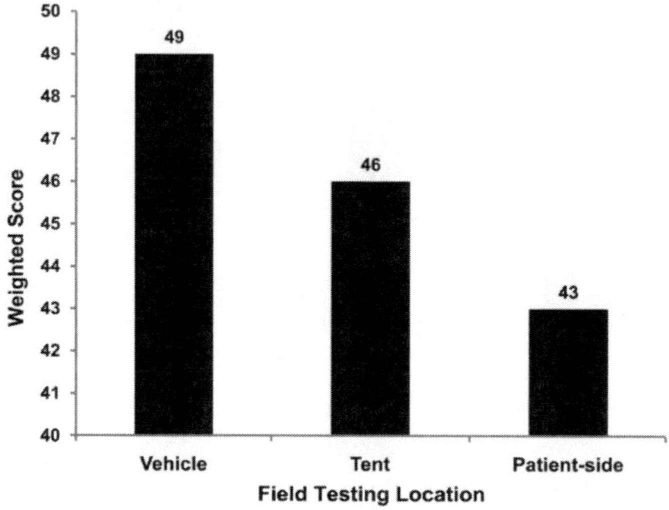

Figure 12. Field Testing Locations. Phang Nga health professionals favored a mobile laboratory concept for field POC testing. However, differences in preferences versus a structure (i.e., tent) placed in the field setting and patient-side testing at the point of rescue or in an alternate care facility were not statistically significant.

Table 7. Top Ranked Diagnostic Test Groups for Near-Patient and Bedside Testing within an Alternate Care Facility

For Near-Patient Testing	Weighted Scores	Rank	Weighted Scores	For Bedside Testing
Pulse oximetry	80	1	63	Blood gases
Chemistry/Electrolytes	58	2	57	Pulse oximetry
Blood Bank/Transfusion	43	3	52	Chemistry/Electrolytes
Cardiac Biomarkers	36	4	29	Cardiac Biomarkers
Blood gases	35	5	28	Hematology
Hematology	17	6	26	Rapid Microbiology Tests
Rapid Microbiology Tests	10	7	24	Coagulation
Coagulation	8	8	9	Blood Bank/Transfusion

Device Design. Table 6 summarizes preferred sample collection methods during disasters by comparing a Vacutainer and hub sample draw using a butterfly needle versus a test cassette that connects by luer lock with also a butterfly needle [8]. Table 6 also shows preferences regarding a test cassette that connects by luer lock with a standard butterfly needle for coupled sample collection versus a built-in retractable needle for direct sample collection [8]. Respondents evaluated four important criteria including lowest risks of contamination, operator error, and biohazard disposal, as well as lowest recurring cost. In their overall assessment, respondents (N = 23) preferred a test cassette sample (score 62) over a Vacutainer and hub (53) for disaster settings. For the comparison of coupled and direct sample collection, respondents (N = 21) selected direct (70) over coupled samplings (35) in disasters. While results were not statistically significant, respondents preferred mobile POC testing from vehicles (Figure 12).

Near-patient and Bedside Testing within an Alternate Care Facility. Survey respondents selected pulse oximetry (score 80), chemistry/electrolytes (58), blood bank/transfusion (43), and cardiac biomarkers (36) as first through fourth choices for near-patient testing (Table 7) [8]. While for bedside testing, they choose blood gases (63), pulse oximetry (57), chemistry/electrolytes (52), and cardiac biomarkers (29).

CONCLUSION

To our knowledge this is the first focused work that describes how to use POC testing in disaster preparedness to improve future resilience through demographic spatial digital terrain analysis of a small-world network in a low-resource setting. Modern crises, in the form of devastating newdemics [9], appear uniquely and suddenly to affect broad populations and geographic areas, some of which are more vulnerable to catastrophic economic set-backs and loss of human life than others. By means of digital elevation contours, Figure 13 shows unequivocally how costal areas of Phang Nga Province were, and in fact still are, at extremely high risk of direct tsunami damage and incapacitation from flooding, because of extensive low-lying coastal areas. In these topographies and other similar coastal areas at risk, demographic geospatial preparedness depends largely on self-reliance and independence, both inherent characteristics of POC practitioners who receive victims and care for patients in primary care structures (see Figure 2) and small community hospitals, typically with compact emergency and critical care functions located close to emergency entrances (Figure 14).

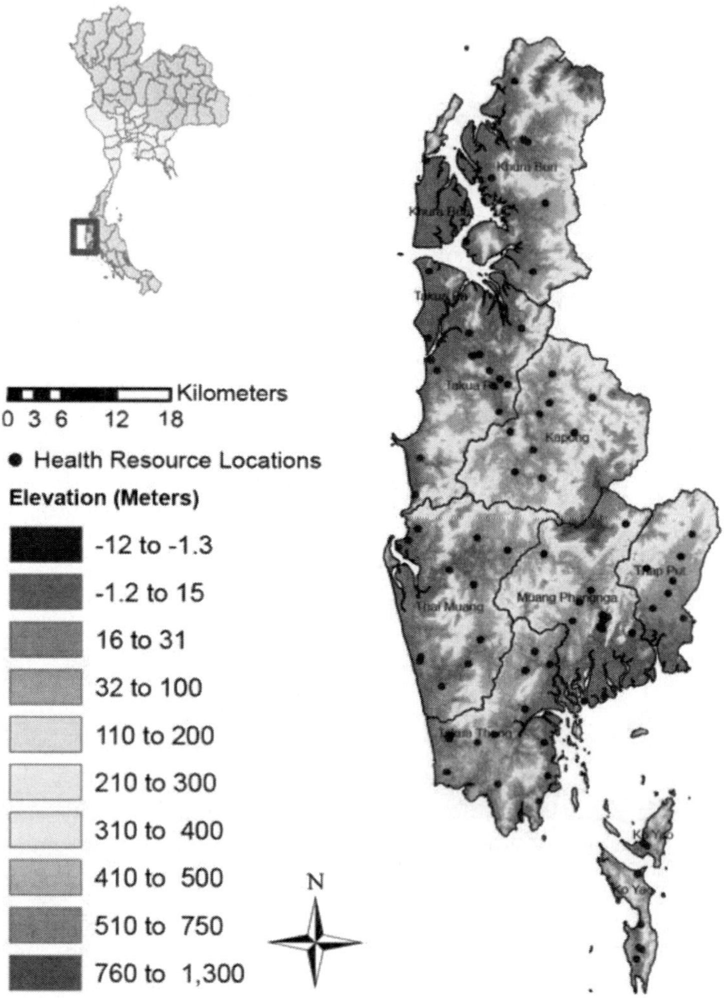

Figure 13. Spatial Preparedness in Phang Nga Province, Thailand. The digital terrain elevations on this topographic map show clearly the vulnerability of health facilities to inundation near coastal areas. The POCT•CTR is developing a POC-SWN-geographic information system (GIS) approach that will facilitate the placement of POC resources. Then, responders will not be stranded without the capability of evidence-based decision-making in the event of another Tsunami or other natural disaster that floods the lowest elevations. A commercial GIS interactive CD also is available to locate health care facilities throughout Thailand from ThinkNet (see http://www.thinknet.co.th/).

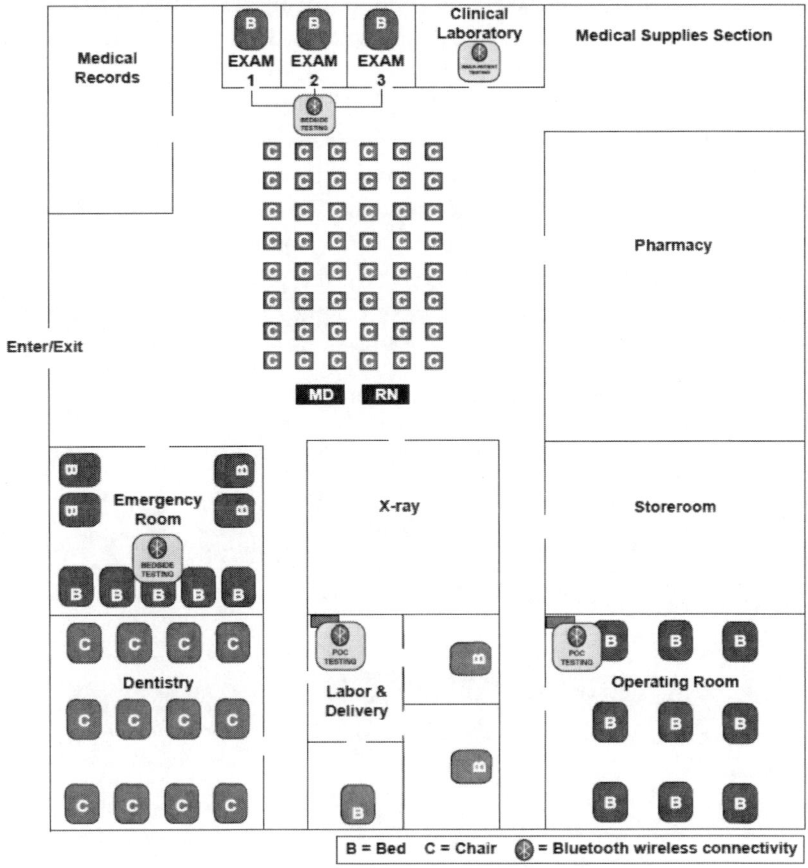

Figure 14. Layout of a Community Hospital. The digital terrain elevations on this topographic map show clearly the vulnerability of health facilities to inundation near coastal areas. The POCT•CTR is developing a POC-SWN-geographic information system (GIS) approach that will facilitate the placement of POC resources. Then, responders will not be stranded without the capability of evidence-based decision-making in the event of another Tsunami or other natural disaster that floods the lowest elevations. A commercial GIS interactive CD also is available to locate health care facilities throughout Thailand from ThinkNet (see http://www.thinknet.co.th/).

Mobile POC testing serves well as a common denominator approach for gathering medical evidence acutely, and by so doing, can enhance crisis standards of care [11], mitigate financial loses, and save lives, or at least reduce morbidity. Ideally, POC resources should not be derived exclusively or even mostly from rescue team caches and external good will, but instead, with planning, be part and parcel of daily routine, urgent, and emergency medical and surgical care. Then, bedside diagnostic evidence obtained with POC instruments can efficiently and quickly avoid deviations in the standard of care in unique circumstances, such as the rational selection of alternate antibiotics based on pathogen identity when first line drugs become exhausted in sites isolated by disasters, or more routinely, evidence-based decision-making based on POC results, such rapid cardiac troponin T and other cardiac biomarker results obtained with the handheld Cobas 232 (Roche Diagnostics, Bangkok), which has been implemented extensively in Thai provinces over the past few years following demonstration of practical need in remote small-world networks along the Mekong River in rural Isaan, northeastern Thailand [10].

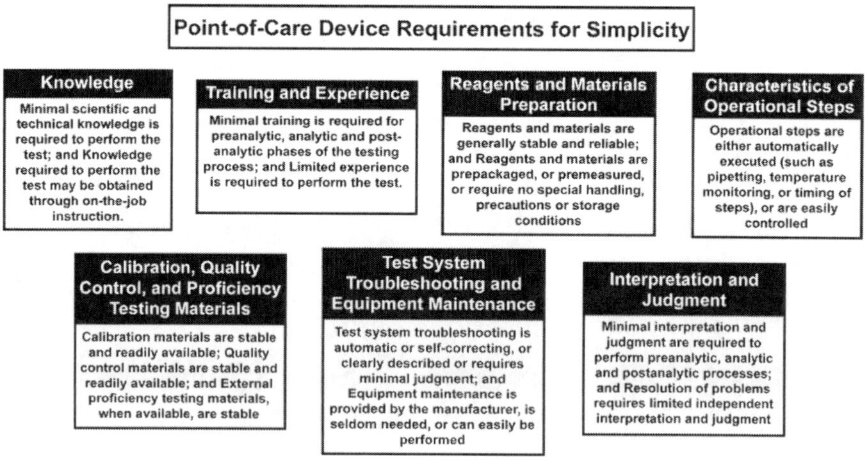

Figure 15. Characteristics of Low Complexity Point-of-Care Devices Suitable for Implementation. Based on the analogy of so-called "CLIA waiver" in the United States, low complexity devices for use by emergency personnel worldwide should embrace the concept of "simplicity" shown categorically in this summary chart. Emergency personnel comprise disaster medical assistance teams, radio-dispatched ambulance drivers, first responders, and possibly even the victims and patients themselves, if they are prepared properly for self-care in a resilient small-world network. CLIA stands for Clinical Laboratory Improvement Amendments of 1988—see http://en.wikipedia.org/wiki/ Clinical_Laboratory_Improvement_Amendments.

Medical leadership can take advantage of readily available resources, such as commercial geographic information systems (GIS) based on interactive CDs that are available from ThinkNet (see http://www.thinknet.co.th/) for Thailand. Geographic information systems can help to spatially locate personnel and integrate health care facilities throughout Phangn Nga Province in order to pre-position adequate resources. This strategy will facilitate the availability of well trained and experienced hospital and emergency staff who know how, when, and why to use POC testing for rapid evidence-based decision-making at the points of need, both on a daily basis to speed care paths and improve cost-effectiveness, and at the time of a future tsunami disaster, should one occur. At the same time, POC devices will grow smaller, smarter, and safer, as well as connected and simpler to use (Figure 15) [24]. Point of care culture [25] inevitably will encompass patients who are empowered to take care of themselves, the ultimate embodiment, expression, and benefit of resilience. We close with recommendations the integrate needs assessment and spatial preparedness.

Evidence-Based Recommendations

1. Reduce Therapeutic Turnaround Time (TTAT) to Speed Critical Paths

- Evidence-based research results show that POC testing decreases therapeutic turnaround time [26,27]. Fast therapeutic turnaround time will improve field response, patient triage, emergency treatment, and alternate care facility operations, which observers [28] deemed important during the Thai Tsunami response.
- Thus, rapid therapeutic turnaround time represents an essential element of small-world network preparedness and a metric by which diagnostic testing effectiveness can be measured.

2. Optimize The Use Of POC Testing

- Routine daily use of POC testing assures high quality and trained POC operators who gain practical experience. Implementation of POC testing should match a) population density (e.g., see Figure 3); b) availability of health care delivery resources, which can be determined by quantitative scoring systems [29]; c) medical needs for simultaneous disaster preparedness and emergency care; and d) geographic proximity to coastal or other regions that risk isolation from flooding and other sequelae of natural disasters.

3. Monitor Continuously In Vivo For Emergency Care

- Cost-effective oxygen saturation monitoring derives from reduced O2 consumption during ambulance transport, especially long distances and durations [30,31], and then, from efficient triage [32] and monitoring [33] in the ED. New designs work well during ambulance transport [34] and can screen patients for carbon monoxide exposure [35].
- Continuous hemoglobin monitoring [36,37] should be explored for decision making during the care of critically ill patients, such as those with Dengue hemorrhagic fever, when hospitalized and in need of frequent hematocrit (hemoglobin) determinations and blood transfusions to keep up with hemolysis.

4. Use Cardiac Biomarkers to Diagnose Acute Coronary Syndromes

- Based on experience and efficacy in low-resource Isaan near the Mekong River [10], POC cardiac biomarker testing can be implemented in community hospital emergency departments and simultaneously improve disaster readiness.
- Diagnosis of acute coronary syndromes using semi-quantitative POC cTn tests (e.g., cTnT, Cobas h232, Roche Diagnostics) will reduce risk when referral sites become overloaded or inaccessible and patients with acute myocardial infarction can receive local treatment promptly.

5. Facilitate Acute and Chronic Diabetes Monitoring in Disasters and Afterward

- Proven effective for diabetes care in other provinces [38], capillary glucose meter monitoring, POC HbA1c testing, and urinary albumin to creatinine ratio (ACR) screening can be implemented in PCUs and community hospitals. However, only high performance POC devices (e.g., DCA Vantage, Siemens Corporation) should be used.

- Appropriate reagents for quality control and patient testing must be available to avoid excess morbidity and mortality when supplies run out during and following disasters.

6. Introduce New Infectious Diseases Testing

- When commercially available, new POC technologies for infectious diseases, such as multiplex pathogen detection by miniaturized and automated nucleic acid testing, can enhance transfusion safety and enable emergency blood donor screening during crises. New molecular pathogen detection devices will be useful in the diagnosis of sepsis in CHs where blood culture and resistance testing are not available.

7. Strive for Geospatial Preparedness

- In regard to POC testing, experience with disasters [39] has shown very clearly that in advance of crises POC coordinators, who provide valuable leadership and strategic planning, should a) know the locations of trained personnel and POC resources, b) have a written disaster plan, c) rehearse operations in realistic simulations of disasters, and d) provide a plan for assuring the quality of diagnostic test results.
- Therefore, it is logical to build, use, and optimize geospatial tools, such as loaded geographic information systems and small-world network knowledge maps, as part of preparedness.

ACKNOWLEDGMENTS

We are indebted to the staff of the Thai hospitals we visited for their courtesy and helpful input during extensive follow-up. We appreciate the support of the College of Populations Studies, Chulalongkorn University,

Bangkok, Thailand. We thank Audhaiwan Suwanyangyuen, MBA, for her diligence when working as a Research Assistant at Chula. This article was supported in part by a Fulbright Scholar Award (Dr. Kost); an Outreach and International Programs grant (Dr. Kost, PI) from the University of California, Davis; the Point-of-Care Testing Center for Teaching and Research, POCT•CTR (Dr. Kost, Director), School of Medicine, UC Davis; and the Edward W. and Ora I. Kost Foundation. Minor support was derived from a National Institute for Biomedical Imaging and Bioengineering (NIBIB) Point-of-Care Technologies Center grant (Dr. Kost, PI, NIH U54 EB007959). The content is solely the responsibility of the authors and does not necessarily represent the official views of the NIBIB or the National Institutes of Health. Portions of the text were modified with permission of the authors (POCT•CTR team) and editors (Dr. Kost and Mr. Curtis) from papers [7,8] that appeared in special issues of the journal, Point of Care, in 2012 and 2013. Figures, tables, and photos (taken by Dr. Kost) were provided courtesy and permission of Knowledge Optimization,® Davis, CA.

REFERENCES

[1] Marti RC. The 2004 hurricanes in the Caribbean and the Tsunami in the Indian Ocean. *United Nations Report (ISBN 92-1-121545-5)*, August, 2005.

[2] 2004 Indian Ocean earthquake and tsunami. *http://en.wikipedia.org /wiki/2004 _Indian_Ocean_earthquake_and_tsunami.* Accessed March 19, 2013.

[3] Banerjee P, Pollitz FF, Burgmann R. The size and duration of the Sumatra-Andaman earthquake from far-field static offsets. *Science.* 2005;308:1769-1772.

[4] Marris E. Inadequate warning system left Asia at the mercy of tsunami. *Nature.* 2005;433(7021):3-5.

[5] Kost GJ. Point of care testing in province hospitals and PCUs: Optimizing critical care and disaster response. In: Charuruks N, Ed. *Point of Care Testing for Thailand.* Bankok: Thai Phim, Chapter 10, pages 159-177, 2007.

[6] Kost GJ, Tran NK, Tuntideelert M, et al. Katrina, the tsunami, and point-of-care testing: Optimizing rapid response diagnosis in disasters. *Am J Clin Pathol.* 2006;26:513-520.

[7] Kost GJ, Katip P, Kulrattanamaneeporn S, et al. Point-of-care testing value proposition for disaster preparedness in small-world networks: Post-tsunami Phang Nga Province, coastal Thailand. *Point Care*. 2013;12:9-22.

[8] Kost GJ, Katip P, Curtis CM. Strategic point-of-care requirements of hospitals and public health for preparedness in regions at risk. *Point Care*. 2012;11:114-118.

[9] Kost GJ. Newdemics, public health, small-world networks, and point-of-care testing. *Point of Care*. 2006;5:138-144.

[10] Kost, GJ, Kost LE, Suwanyangyuen A, et al. Emergency cardiac biomarkers and point-of-care testing: Optimizing acute coronary syndrome care using small-world networks in rural settings. *Point Care*. 2010;9:53-64.

[11] Kost GJ, Sakaguchi A, Curtis C, et al. Enhancing crisis standards of care using innovative point-of-care testing. *Amer J Disaster Med*. 2011;6:351-368.

[12] Wattanawaitunechai C, Peacock SJ, Jitpratoom P. Tsunami in Thailand—disaster management in a district hospital. *N Engl J Med*. 2005;352:962-964.

[13] Kulrattanamaneeporn S, Tuntideelert M, Kost GJ. Using telemedicine with point-of-care testing to optimize health care delivery in Thailand. *Point of Care*. 2006;5:160-163.

[14] Kulrattanamaneeporn S, Wongboonsin K, Kost GJ. Impact of point-of-care testing and telemedicine on diabetes management in primary care unit settings in rural Thailand. *Point of Care*. 2009;8:77-81.

[15] Kost GJ, Katip P, Vansith K, et al. The final frontier for point of care: Performance, resilience, and culture. *Point of Care*. 2013:12:1-8.

[16] Kost GJ, Katip P, Vinitwatanakhun C. Diagnostic testing strategies for healthcare delivery during the Great Bangkok Flood and other weather disasters. *Point of Care*. 2012;11:191-199.

[17] Kost GJ. Theory, principles, and practice of optimizing point-of-care small-world networks. *Point of Care*. 2012:11:96-101.

[18] Bureau of Policy and Strategy, Ministry of Public Health. Mid-year population data in 2010 of Phang Nga Province Available at *http://bps.ops.moph.go.th/ index.php?mod=bps&doc=5_1*. Accessed March 19, 2013.

[19] Bureau of Policy and Strategy, Ministry of Public Health. Numbers of medical personnel in community and regional hospitals classified by provinces of 2010. Tables 17 and 18. Available at *http://hrm.moph.go.th/*

res53/res-rep2553.html. Accessed March 14, 2010. [Also 2009 for medical technologist data.]

[20] Bureau of Policy and Strategy, Ministry of Public Health. Numbers of regional and community hospitals and hospital beds classified by regions and provinces of 2010. Table 3. Available at *http://hrm.moph.go.th/res53/res-rep2553.html.* Accessed March 14, 2010.

[21] The numbers of deaths, injuries, missing following earthquake and Tsunami in Phang Nga Province—Data of January 10, 2005, 18.00 PM. Available at *http://www.photoontour.com/events_html/tsunami/data/data04.htm.* Accessed March 19, 2013.

[22] National Statistical Office, Prime Minister's Office. Map of districts (amphurs), sub-districts (tambons), and municipal areas, and fundamental data of Thai provinces in 2000. *Bangkok: Data and Information Technology Division, National Statistical Office*, 2000; pp.13-332 (in the Thai language).

[23] Jaiyen W. *Distances and travel times between primary care units and community hospitals and between community and regional (referral) hospitals in Phang Nga Province.* E-mail correspondence, November 18, 2010.

[24] Curtis CM, Kost GJ, Louie RF, et al. Point-of-care hematology and coagulation testing in primary, rural emergency, and disaster care scenarios. *Point Care*. 2012;11:140-145.

[25] Kost GJ, Katip P, Vansith K, et al. The final frontier for point of care: Performance, resilience, and culture. *Point Care*. 2013;12:1-8.

[26] Kilgore ML, Steindel SJ, Smith JA. Evaluating stat testing options in an academic health center: therapeutic turnaround time and staff satisfaction. *Clin Chem*. 1998;44:1597-1603.

[27] Kost GJ. The Hybrid Laboratory, Therapeutic Turnaround Time, Critical Limits, Performance Maps, and Knowledge Optimization.® In: Kost GJ, Editor. *Principles and Practice of Pont-of-Care Testing*. Philadelphia: Lippincott Williams & Wilkins, 2002, chapter 2, pages 13-25.

[28] Leiba A, Ashkenasi I, Nakash G, et al. Response of Thai hospitals to the tsunami disaster. *Prehosp Disaster Med*. 2006;21:32-37.

[29] Kost GJ, Katip P, Kanokslip A, et al. A new demographic strategy for point-of-need medical testing: Linking health resource scores, poverty levels, and care paths. *J Demography* (Chulalongkorn University, Bangkok). 2011;27:1-31.

[30] Macnab AJ, Susak L, Gagnon FA, et al. The cost-benefit of pulse-oximeter use in the prehospital environment. *Prehosp Disaster Med.* 1999;14:245-250.

[31] Howes DW, Field B, Leary T, et al. Justification of pulse oximeter costs fro parametric prehospital providers. *Prehosp Emerg Care.* 2000;4:1515-155.

[32] Summers RL, Anders RM, Woodward LH, et al. Effect of routine pulse oximetry measurements on ED triage classification. *Am J Emer Med.* 1998;16:5-7.

[33] Jones J, Heiselman D, Cannon L, et al. Continuous emergency department monitoring of arterial saturation in adult patients with respiratory distress. *Ann Emerg Med.* 1988;17:463-468.

[34] Weber U, Tomschik E, Resch I, et al. Comparison of two new generation pulse oximeters during emergency ambulance transportation. *Eur J Anaesthesiol.* 2011;28:92-96.

[35] Nilson D, Partridge R, Suner S, et al. Non-invasive carboxyhemoglobin monitoring: screening emergency medical services patients for carbon monoxide exposure. *Prehosp Disaster Med.* 2010;25:253-256.

[36] Kost GJ, Tran NK. Continuous noninvasive hemoglobin monitoring: The standard of care and future impact. *Crit Care Med.* 2011:39:2369-2371.

[37] Frascas D, Dahyot-Fizelier C, Catherine K, et al. Accuracy of a continuous noninvasive hemoglobin monitor in intensive care unit patients. *Crit Care Med.* 2011;39:2277-2282.

[38] Kost GJ, Kanoksilp A, Mecozzi DM, et al. Point-of-need hemoglobin A_{1c} for evidence-based diabetes care in rural small-world networks: Kumuang Community Hospital, Buriram, Thailand. *Point Care.* 2011;10:28-33.

[39] Mann P, Kost GJ. Point-of-care coordinator leadership in preparedness, crisis, and recovery. *Point of Care.* 2012;11:102-107.

In: Tsunamis
Editor: Tianxing Cai
ISBN: 978-1-62808-682-9

Chapter 2

ADAPTIVE CAPACITY BUILDING THROUGH PUBLIC PARTICIPATION: A BOTTOM-UP EMPOWERMENT PROJECT IN THE COASTAL SUBSIDED AREA OF TAIWAN

Sinite Yu[1], Kate Yeong-Tsyr Wang[2,*], Hong-Yuan Lee[3], Bi-Liang Lin[4] and Wen-Yao Hsu[5]

[1]Taiwan International Institute for Water Education
[2]National Taiwan Normal University
[3]National Taiwan University
[4]Bread of Life Christian Church in Dongshi
[5]Taiwan International Institute for Water Education

ABSTRACT

The studied area has been identified as the first climate refugees on the west coast of Taiwan. Due to the groundwater over-pumping for the industry usage, the land subsidence has been observed during the past 20 years. With the global climate change, the phenomenon of sea level rising has worsened the problem of flooding in that area. Besides land subsidence and flooding, water shortage, water pollution, industrial

* Corresponding Author address: Graduate Institute of Social Work,National Taiwan Normal University,No.162, Sec 1,Be-ping East Road,Taipei 10610, Taiwan, Email: ytwang@ntnu.edu.tw.

decline, destroyed public intra-structure and the migration of young people to cities are also the critical problems in the area. Therefore, how to build the adaptive capacity of the residents to communicate with the government and corporations for enhancing the well being of the locals becomes a pivotal issue in the recent years.

This project has started since 2006 and the action research approach has been used. Since 2006-2008, local major and influential stakeholders were interviewed for a better understanding of the problems people faced in that area. From 2009-2011, different types of community activities and classes have been implemented. The locals learned the related knowledge and express their own opinions through the above channels. In order to contact the most vulnerable people who are the poorest group, the face-to-face interviews were conducted at their houses. This out-reach approach helped the connection with those who seldom make their voice heard in public. The third phase begins in 2012. A community association in a fishing village was organized to strengthen the long-term relationship between the locals and the project team. Through more comprehensive community contacts and activities, it is believed that the locals' adaptive capacity can be built up and the future directions of community development can be pursued. Four types of the community activities were held during these years and there were interviews, group discussions, workshops and conferences. This bottom-up empowerment project not only provides various communication channels but also raises the awareness of climate changes of the locals. More actions will be taken for reaching the seven consensus goals during the third phase of the project. To sum up, this paper will explore the bottom-up, public participation approach for confronting the impacts of the global climate change.

INTRODUCTION

Citizens of fishing villages in western Taiwan are the first group of climate refugees of Taiwan. In addition to land subsidence over time, they face frequent floods from rapid sea level rises as the increasing frequency of heavy rainfall due to global climate change.

This study investigated how professional teams and local communities raise a new capability to resolve environmental issues relating to climate change through public participation by examining the case of the Dongshi Township in Chiayi County of Taiwan. As a bottom-up power, public participation can clearly identify integrated complex issues such as industrial decline and social marginalization to reach a consensus for national land restoration and thereby assist the government in suggesting solutions through related policies.

Bottom-up public participation in decision-making and empowerment is a new dynamic strength for resolving environmental issues related to climate change. Starting from 2006, the field project was divided into three periods: Exploration Period 2006-2008, Empowerment Period 2009-2011, and Long-Term Community work Period started from 2012.

The past six-year efforts have laid the foundation for developing public participation opportunities. However, there are some limitations to public participation, as the procedure and method of public participation are unorganized, and the public is not delegated with substantial decision-making authority. Public participation in decision-making and empowerment is a slow process. Although public participation began in 2006, it does not mean that communities now have adequate ability to resist the traditional, handicapped policies. More efforts are still needed for strengthening locals' abilities in public participation. The following sections will introduce the environmental, social and economic contexts of Dongshi Township, and then present how the public participation approach had been implemented over the years. The future challenges will also be discussed in the final section.

Environmental Overview of Dongshi Township in Chiayi

With an area of 1,901.6725 km2, Chiayi County occupies 5.35% of Taiwan's total area. Dongshi Township is located in the western Chiayi County, with an area of about 82 km2. At the westernmost point of Chiayi, Dongshi is situated with the Taiwan Strait to the west, the Beigang River and Kouhu Township in Yunlin County to the north, Budai Town to the south, and Puzi City to the east (Figure 1). The Puzi River runs across it, and the entire township is located in a flatland area. There are levees and seawalls to the north and west protecting the township from tidal and flood effects. The ground elevation of Dongshi Township is less than 10m. The local terrain slopes gently, and part of Dongshi is below the sea level. Dongshi has a 14km-long coastline, and the coast is flat and shallow. As there are many sandbanks in the township, locals build oyster racks on these sandbanks to cultured oysters.

Although there are very limited surface water resources in coastal Chiayi, water demand has been increasing in recent years due to rapid agricultural and industrial development. Due to groundwater over-pumping, land subsidence has occurred. Dongshi Township and Budai Town are the center of subsidence and account for the land subsidence in Chiayi County. Land subsidence began in Chiayi in the 1970s. During that period, the groundwater table in coastal Dongshi was about 10m below sea level; and by 1996, it had sunk in more than 24m below sea level.

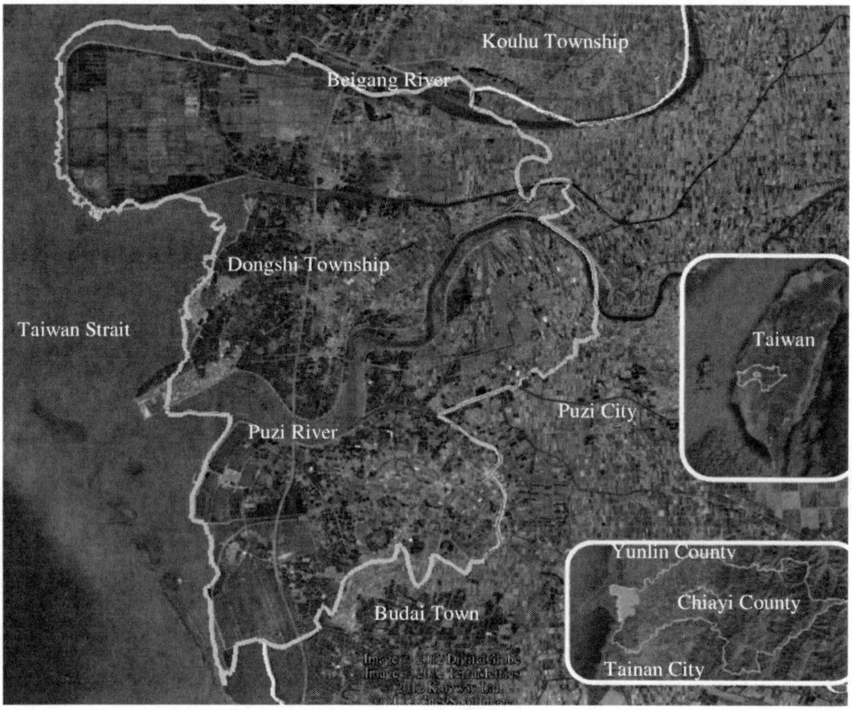

Figure 1. Location of Dongshi Township in Chiayi County.

LAND SUBSIDENCE

Dongshi Township is the center of accumulated subsidence in Chiayi County. In 1988-1999, the subsidence level of Sanjia Village was 1.17m; in 1992-1993, the average subsidence rate was 15cm/yr; and the average subsidence began to decline after 1997. From 1988 to 2009 the accumulated subsidence of Sanjia Village was 1.42m[1]. By the record of last 30 years, the land subsidence of Dongshi has reached more than 2 meters.

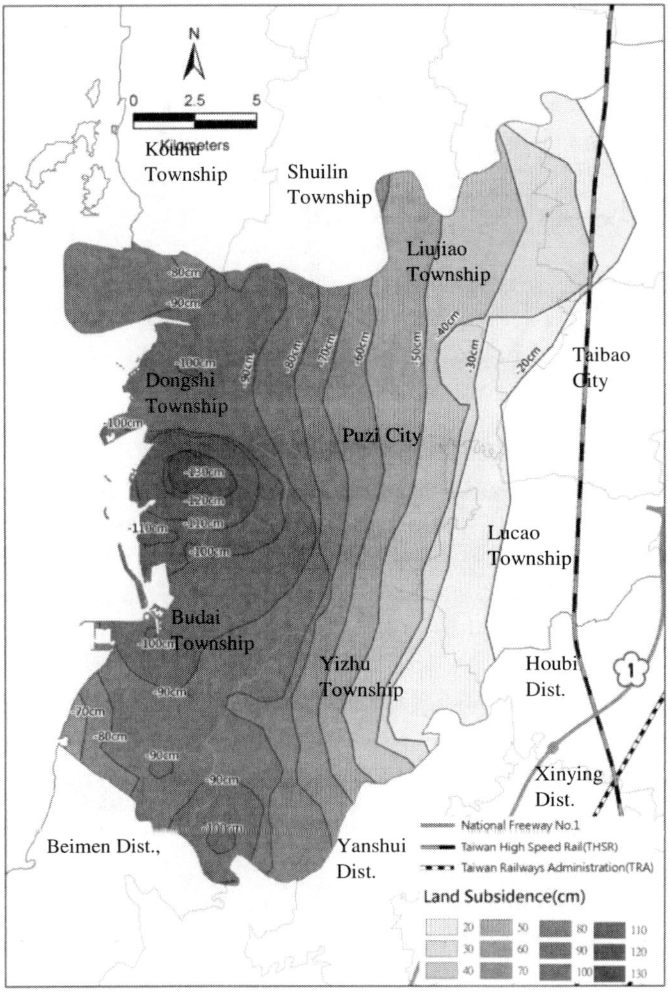

Figure 2. Accumulated Subsidence in Chiayi during 1991-2012.

1991 to 2004 was the period with the most significant land subsidence in Chiayi County. The area with more than 90cm of accumulated subsidence

[1] San-jiang Police Station monitoring point in Sanjia Village, Land Subsidence Database of Taiwan. (Available online: http://www.subsidence.org.tw)

included Wanliao Village and Wenzhi Village in Dongshi Township; and more than 80% of Dongshi Township experienced land subsidence of more than 80cm. The distribution of land subsidence during 1991-2012[2] is shown in Figure 2.

According to the 2006 statistics, there were 2,692 groundwater pumping wells in Dongshi Township, with estimated annual pumpage about 15,720,000 m^3; including 1,154 wells for irrigation and 1,475 wells for aquaculture, which were the major sources of pumpage at about 1,370,000m^3 and 13,490,000 m^3 respectively. The annual pumpage of aquaculture was about 85.8% of the total annual pumpage known in the area.

FLOODS IN LOW-LYING AREA

As a result of land subsidence, about 2,250 hectares of land in Chiayi are now below sea level, and about 7,350 hectares of land area below the mean tide level. If heavy rains and storm surges come at the same time, severe floods occur and inconvenience and damage daily life and business activities. SSHS Cat. 5 typhoons that caused severe floods in the coastal area included Herb in 1996 and Mindulle in 2004. On August 1, 1996, Herb brought 772.7 mm of rainfall to Chiayi. Due to seawater invasion, for several days nearly all of Dongshi Township and Budai Town were inundated[3]. On June 30, 2004, Mindulle brought more than 400-500 mm rainfall to central and southern Taiwan. Along with seawater invasion, Changhua, Yunlin, and Chiayi counties were all flooded.

On June 12, 2005, Aogu in Dongshi Township, Chiayi County, recorded 102 mm rainfall in one hour. As a result, low-lying areas were severely flooded, and part of Dongshi Township was inundated for six days. When torrential rainfall met the high tide on June 9, 2006, Dongshi Township was inundated again as it was difficult to drain flood water in downstream areas. When typhoon Morakot hit Taiwan on August 8, 2009, it brought 2,776.5 mm of rainfall to Chiayi in three days and traumatized central and southern Taiwan[4].

As these floods were catastrophic, the Water Resources Agency implemented the Enforcement Land Subsidence Area Drainage Environment

---

[2] Land Subsidence Database of Taiwan. (Available online: http://www.subsidence.org.tw)

[3] Public Construction Commission, Executive Yuan (1998) The Typhoon Herb disasters and reconstruction engineering documentary.

[4] Taiwan International Institute for Water Education (2011):2-7~8.

Improvement Demonstration Project to protect villages with flood defense facilities. However, as some villages did not agree with the land use change and the form of facilities, some projects were unfulfilled. Even though all the projects were completed, they were still unable to resolve local flood problems as a result of land subsidence and climate change.

INDUSTRIAL DECLINE AND SOCIAL MARGINALIZATION

The distribution of the fishery and agricultural populations in Dongshi Township is shown in Figure 3. The local fishery population is about 10,000 people, with no significant change. By contrast, the local agricultural population has declined significantly since 2005[5].

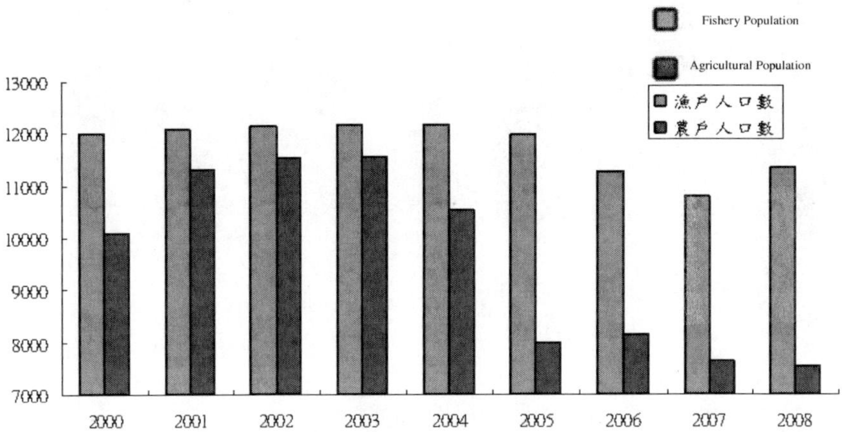

Data source: Statistics of the Accounting and Statistics Department of Chiayi County Government

Figure 3. Distributions of Fishery and Agricultural Populations in Dongshi Township 2000-2008.

The total aquaculture area in Dongshi Township is 3,670.47 hectares, with aquaculture occupying an area of 2,060 hectares (56.12% of the total) in shallow sea. In inland culture, saltwater fish ponds occupy a larger area about 1,060.29 hectares (28.89% of the total), and retired fish ponds occupy an area of 393.21 hectares (10.71% of the total). Major species of aquaculture include

[5] Statistics of the Accounting and Statistics Department of Chiayi County Government.

clams (intermixing with milkfish), perch, sea bream, eel, and shrimp. Pond owners usually culture fish with low-salinity water or fresh water. Wengang Village became the center of oyster supply under the name of "Dongshi Oyster", and oyster shell splitting is the work of most local women[6] (Figure 4).

Fish ponds in Dongshi Township were badly damaged by floods, and pond owners suffered severe damage from the loss or death of cultural products. As the township is often inundated with heavy rain, the local living quality is low, and it is thus unable to attract investment in industries other than agriculture. Due to capital shortage, high cost, the unstable price of fish product, data shortage for fishery resource development, and water pollution, the local fisheries require an immediate boost of competitiveness.

Fishermen in a fish pond (© TIIWE)

Oyster shell splitting (© TIIWE)

Figure 4. fishery in Dongshi Township.

As a result of industrial decline, there was an ongoing job shortage, and population outflow began. In 1985, the population of Dongshi Township was 36,912 people; in December 2012, the total population of the township was 26,713 people, with a reduction of 10,199 people more than 27% of population over the past 15 years. The population fall in both number and percentage was exceptionally high between 1990 and 2000 (Table 1).

Low fertility, population outflow, and population aging are the factors affecting social development in Dongshi Township. As shown in Table 2, the working age population (aged 16-64) from 2005 to 2012 declined from 19,669 people to 18,337 people; and the youth population declined from 4,254 people to 2,945 people. Due to low fertility and population ageing, the youth

[6] Data Source: Statistics of the Accounting and Statistics Department of Chiayi County Government.

dependency ratio thus declined from 21.63% to 16.06% and the old age dependency ratio rose from 27.71% to 29.61%.

Table 1. Population Change in Dongshi Township 1980-2012

Year	Population	Population Reduction	Population Reduction Rate
1985	36,912	-	-
1990	33,549	-3,363	-9.11%
2000	29,957	-3,592	-10.71%
2005	29,374	-583	-1.95%
2010	27,741	-1,633	-5.56%
2012	26,713	-1,028	-3.71%

Data source: Statistics of the Accounting and Statistics Department of Chiayi County Government.

Table 2. Population Ageing in Dongshi Township 2005-2012

Age / Year	0-14 (People)	15-64 (People)	65 or older (People)	Old Age Dependency Ratio (%)	Youth Dependency Ratio (%)	Dependency Ratio (%)
2005	4,254	19,669	5,451	27.71	21.63	49.34
2006	4,115	19,374	5,489	28.33	21.24	49.57
2007	3,945	19,275	5,541	28.75	20.47	49.21
2008	3,794	19,018	5,587	29.38	19.95	49.33
2009	3,623	19,011	5,590	29.40	19.06	48.46
2010	3,379	18,832	5,530	29.36	17.94	47.31
2011	3,127	18,622	5,453	29.28	16.79	46.07
2012	2,945	18,337	5,431	29.61	16.06	45.67

Source: Statistics of the Accounting and Statistics Department of Chiayi County Government.

PUBLIC PARTICIPATION

Land subsidence in coastal Chiayi has brought comprehensive problems, such as floods, water resource shortage, water pollution, infrastructure damage, industrial decline, and social marginalization (Figure 5). Along with the threat of climate change, these problems are complicated. It is thus necessary to build a new capacity to resolve these problems.

Bottom-up public participation can be the new momentum in resolving these complex problems in Dongshi Township. Such bottom-up public participation needs professional facilitation to empower local residents capable to ensure dialogue with professionals and the government, so as to promote three-way cooperation for solutions to related problems. Hence, building such new capacity was started in 2006. The process fell into the following three stages: Exploration Period 2006-2008; Empowerment Period 2009-2011; and Long-Term Community work Period from 2012.

Floods in tombs area (© TIIWE)

Oyster shell piles in community (© TIIWE)

Figure 5. Environmental problems in Dongshi Township.

A. Exploration Period 2006-2008

In the exploration period of 2006-2008, local issues and concerns among the major stakeholders through public participation were captured. Public participation was achieved through group interviews, local workshops, and thematic workshops. First, interviews (Table 3) were planned for six themes and conducted on respondents with high data density and strength. After the group interviews, local workshops were held in various villages in the Dongshi Township to discuss integrated issues relating to planning. Lastly, thematic workshops were held based on those six themes. The outcomes of public participation and the consensuses formed were included in the National Land Restoration and Sustainable Development Planning in Dongshi Township in Chiayi County promoted by the government.

B. Empowerment Period 2009-2011

During the empowerment period of 2009-2011, public participation and empowerment courses were administered to educate the stakeholders about empowerment, so as to extend the momentum of public participation. In this period, the Water Resources Agency selected the scope of the demonstration area, and central government departments proposed related plans for the demonstration area in Dongshi Township. According to the planning, land uses for industrial, recreational, wetland and cemetery were planned in the area. During this period, public participation focused on villages with large changes in land use. These included Xingcuo, Yuanshu, and Yongtun villages. Face-to-face interviews were conducted with local families to better understand the needs of these minority groups. Questionnaires were applied to help underrepresented groups discuss their needs.

Table 3. Correspondence between Planning Themes and Interview Themes

Item	Planning Theme	Interview Theme
1	Flood Control and National Land Restoration	Priority of flood control, difficulty in implementation, and keys to successful promotion
2	Water Consumption and Industry	Common damage in culture and agriculture, adjustment of water consumption characteristics, and sustainable culture strategy
3	Ecology and Tourism	Prospects of ecotourism, opportunities for tourism, and planning of sustainable ecotourism strategy
4	Life and Environment	Challenges of living and work in contentment, suggestions for environmental improvement, and maintenance strategy for sustainable environment,
5	Policy and Land	Sustainable costal land use planning
6	Flood Relief, Recovery, and Revitalization	Decision support information

First, three categories of stakeholders were divided: hazard risk victims (including present and future risk), resource providers, and responsibility actors. Four such groups were formed: administration agency representative, industrial representative, technical expert, and the general public; i.e. The government, opinion representatives, officials, industrialists, experts or NGO groups, and hazard victims (particularly minority or underrepresented groups) (Figure 6).

Figure 6. Relational Chart of Project Stakeholder Groups.

Table 4. List of Stakeholder Groups

Stakeholder Group	**Recommend List**
Government	Water Resources Agency, Fisheries Agency, Construction and Planning Agency, Department of Health, Tourism Bureau, Fisheries Research Institute, County Government, Township/ Village Offices, etc.
Experts and Scholars	Civil engineering experts, hydrology experts, land administration experts, and aquaculture experts
Business Sector	Culture farm owners, travel agents, and land developers.
NGOs	Dongshi Boating Association, Chuanzitou Foundation, Chiayi Ecological and Environmental Conservation Association, National Taiwan University Building & Planning Foundation, and Chiayi County Sustainable Development Association
Community Workers	Local community development associations and community planners
Social Welfare agencies	Public and major private social welfare agencies in Chiayi
Local Citizens	Particularly Low-income or near poor families

As it was necessary to understand the intention of the participants before forming the stakeholder groups, related activities were organized to inquire about the intentions of stakeholders. In doing so, related stakeholder groups were formed and personnel or representatives at different levels to discover the issues of different levels through dialogues with these groups were also located. The recommended list is shown in Table 4. The public participation methods in this period included face-to-face interviews, open interviews, group discussions, workshops, and conferences.

The process of various dialoguing activities is diagrammed in Figure 7. The relational analysis between the forms of participation and the theme, scale, interaction strength, and mutual trust strength are shown in Table 5.

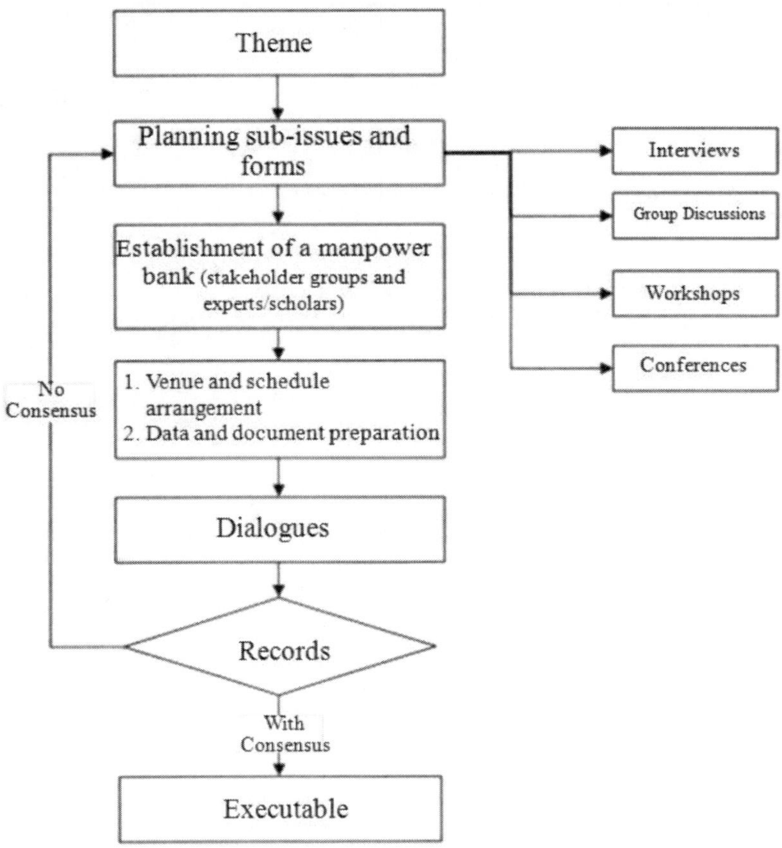

Figure 7. Planning Process of Organization Dialogue.

Table 5. Activity Relations among Different Forms of Dialoging Organizations

Activities / Compared Items	Interview	Group Discussion	Workshop	Conference
Mutual Trust Development	●	●	◎	◎
Interaction	●	●	◎	◎
Scale of Conference • Numbers of Participant • Time needed for Preparation • Information Transparency	○	○	◎	◎
• Thematic • Issue Development	○	○	◎	◎
Time Duration of Activity	1-1.5 hrs	1.5-2.0 hrs	2.0-3.0 hrs	3.5-4.0 hrs

Note: ● high strength; ◎ medium strength; ○ weak strength.

The method, participants, and manipulation of dialogues are described below:

1.1. Face-to-Face Interviews of Vulnerable Groups

Local citizens are always the most important role among all stakeholder groups on all issues, and this group usually includes underrepresented groups. This project conducted questionnaire interviews with low-income and near poor families in Xingcuo, Yuanshu, and Yongtun villages in Dongshi Township, Chiayi County.

It was planned to survey 300 families, but only 205 families were surveyed due to list quality, project schedule, and the aforementioned factors. After deducting invalid responses (repeated list, interview rejection, inability to respond due to physical disability and age, and paper households), there were 177 valid responses.

It was a questionnaire interview investigating the flood impact on local citizens, with items falling into the following nine categories, including (1) basic data, including financial status and religion; (2) flood experience, including family flood experience in recent years; (3) flood impact on the schooling of children in the past two years (including nursery, kindergarten to senior high schools); (4) health impact of floods in the past two years; (5) flood impact on living in the past two years; (6) flood impact on participation in community/social groups or activities in the past two years; (7) flood impact on work in the past two years; (8) flood impact on family incomes in the past two years; and (9) flood impact on utilities in the past two years.

1.2. In-Depth Interviews of Other Stakeholders

Another semi-structured questionnaire was designed, with items falling into the following 12 categories: (1) flood impact, (2) flood problem management, (3) social welfare need, (4) social welfare supply and demand, (5) use of social welfare services, (6) discussions on the development of potential community resources, (7) present status and suggestions for flood control projects, (8) present status and future development of aquaculture, (9) land acquisition problems, (10) allocation of interest, (11) industrial development, and (12) process and suggestions for public participation.

Respondents of the second questionnaire included local citizens, community staffs, community representatives, and focus groups. Some respondents were employees of NGOs with a long history in local areas and a good understanding of local conditions. A total of 46 interviewees interviewed on flood control, disaster prevention, and the living environment, with a total time of participation of 56.5 hours, in other words, 1.64 hours and 3.98 respondents for each session. The outcomes of these interviews suggest that policy planners should try to better understand the living conditions of citizens living in flooded areas, so as to better consider the needs of citizens when planning flood control policies, and to develop comprehensive policy-making solutions, instead of dealing with problems on an ad hoc basis.

2. Group Discussions

Through the process and based on the outcomes of face-to-face interviews with minority groups and open interviews, the role of the professional team was to continuously extend the scope of interviews to all parts of the demonstration area, so as establishing a network list with broad coverage, discovering influential and enthusiastic opinion leaders or people with deep experience in local development, and inviting them to participate in all forms of public participation activities planned in the future. Group discussion participants included local government employees, heads of village, school head persons or principals, teachers, and enthusiastic opinion leaders. (Figure 8) A total of 96 people participated in 11 group discussions for a total of 23 hours, or nine participants for each session in average, based on the following subjects: "flood control and disaster prevention", "living environment", "flood control projects and interview: a social worker perspective", "local volunteer force empowerment", "sharing of recent flood control experience", "environmental protection volunteer survey", "visions of present flood control", "present policies of the Fisheries Agency in addressing land subsidence in coastal Chiayi", "village flood control strategies", and "opportunities community empowerment and rural renaissance".

Figure 8. group discussion (© TIIWE).

3. Workshops

The outcomes of interviews and group discussions became the bases of the workshops, and workshop issues were determined through preliminary opinion extraction. The scale of the workshops differed from that of the group discussions as there were clearer agenda. The focus of issues was gradually narrowed to the coexistence of ecology, living, and production. These issues were coherent and continuous in public participation, from the interviews and seminars to the workshops. Also, the depth of investigation increased.

There were two types of workshops: manpower empowerment and thematic discussion workshops. The former aimed at empowering locals, so that they could better understand the knowledge of land subsidence. Also, the locals' feedback on land subsidence was collected through discussions.

Interviewers and supervisors of face-to-face interviews also received empowerment training to improve their understanding of local conditions to help them to conduct subsequent face-to-face interviews.

Based on the nature of the issues discussed at the workshops, local government employees, village heads, community staff, and opinion leaders and key persons from previous public participation events, particularly professionals and experts in related areas, were invited to educate participants about professional knowledge in related areas.

A total of 398 participants participated in 20 workshops for a total of 56.5 hours, or 20 people for each session (Figure 9), and held based on the following themes: "manpower empowerment training", "flood control and aquaculture", "flood control, disaster prevention, and the living environment", "aquaculture and water consumption", "ecology and tourism", "the application of green living perimeters", "flood control, green life, and rural development", "land use development, townscape, and finance", and "sustainable eco-aquaculture".

4. Conferences

Different types of issues discussed in the interviews, group discussions, and workshops converged on production, living and ecology issues. In order to refine the outcome of public participation, conferences were held to include more members of society and to filter previously collected outcomes before presentation to the stakeholder groups, so as to gather different points of view and thereby discover new ideas.

Conference participants included representatives from central and local governments, village heads, school head persons and principals, teachers, local NGOs representatives, business sector, opinion leaders, and experts. Three conferences were held based on the following themes: "industrial revitalization and life improvement", "flood control and living", and "aquaculture and groundwater". At these conferences, status reports on local conditions were presented, and issues and solutions were discussed. Overall, the outcomes of public participation were presented.

The cohesion process of consensuses of public participation is shown in Figure 10. In practice, comprehensive sub-issues were collated from the flood control opinions collected from extensive public interviews. Seminars and workshops were held to discuss these sub-issues. Lastly, conferences were held to discuss the consensuses gathered from these three forms of public participation.

Figure 9. workshop (© TIIWE).

The true needs and issues of concern to the public were investigated through intensive interviews. These issues were effectively focused on group discussions. Major stakeholder groups were invited to the workshops based on these issues for further communication and interaction. After focusing these issues on three main issues, namely ecology, living, and production, public conferences were held to reach consensus among participants.

With such an action research method, the local development consensus—co-existence of aquaculture and groundwater use ensuring "inundation relieving", "environment interesting", "water supply unfailing", "water quality cleaning", "fisheries assuring", "life relaxing", and "profit keep earning"—was reached as a suggestion for the government to assess the efficacy of the national land recovery demonstration plan for coastal land subsidence areas.

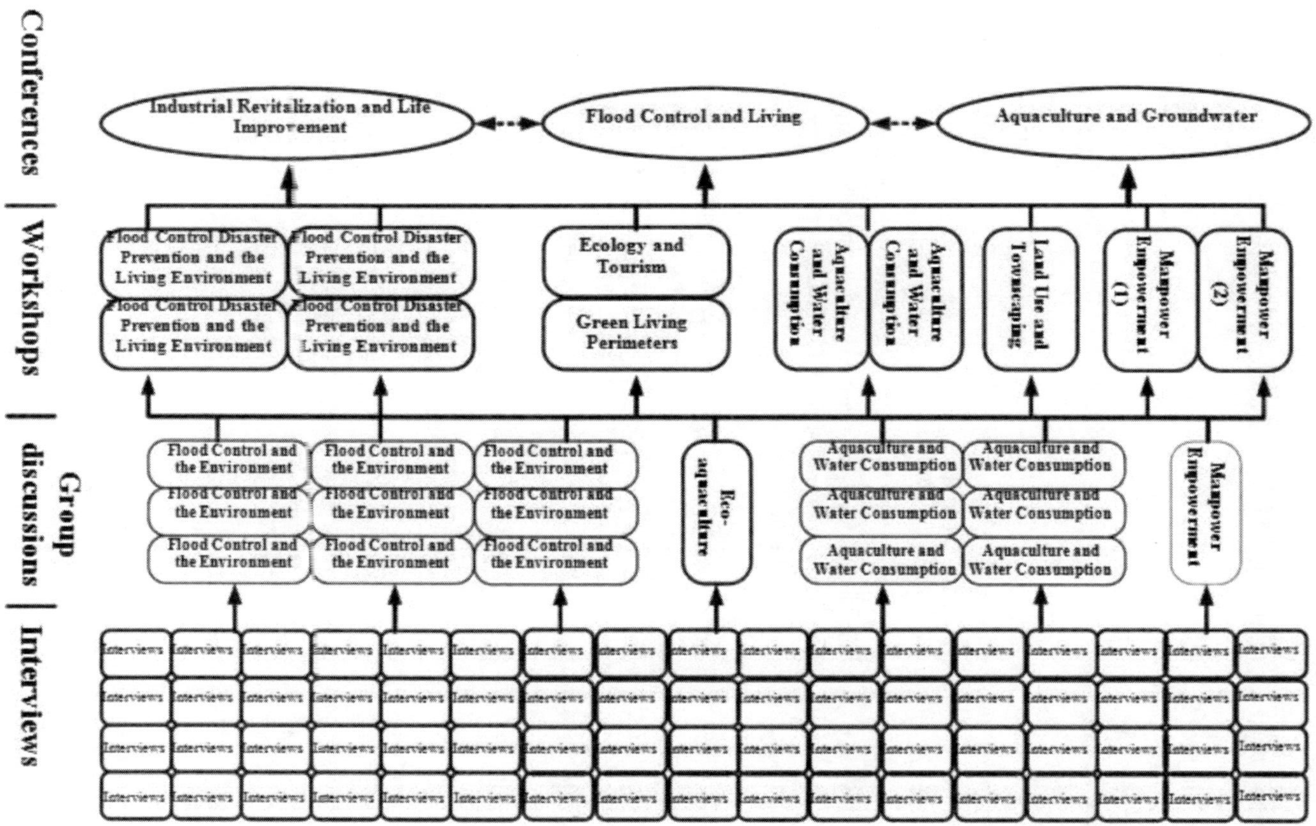

Figure 10. Illustration of Issue Cohesion of Public Participation.

C. Long-Term Community Works Period from 2012

The long-term community residence period began in 2012. From then on, community-concern associations in communities were organized to engage in local social work. Also, a new set of indicators was developed through public participation to assess the government's future administration directions and offer suggestions.

Three workshops (Figure 11) and one seminar were held in this period. Participants of these four events can be categorized into the public sector, experts and scholars, and the volunteer sector. The public sector included agencies from the central government, such as the Council for Economic Planning and Development (CEPD) and Water Resources Planning Institute Agency (WRAP); local agencies including related bureaus and departments of the Chiayi County Government, farmer's and fishermen's associations, and village heads; and the head persons, principals and deans of local elementary and junior high schools. Scholars and experts included professors of the Graduate Institute of Social Work of National Taiwan Normal University; professors of the Department of Water Resources Engineering and Conservation of Feng Chia University; and professors of the Department of Environmental Management of Tungnan University; and hydraulic engineers, and the volunteer sector included community organizations, businesses, and NGOs.

As the integrated planning for local environmental improvement emphasizing ecology, living, and production involves comprehensive aspects, instead of a "trans-regional" or "interdisciplinary" body, a project implementation framework that can clearly understand and practically address local demands and resolve urgent problems in local areas should be established through cooperation and a division of labor between government units of different levels and with different expertise. As consensus on living, local development, and eco-tourism development was reached at this stage of public participation, the government should respond to the following two objectives:

1. Address local demands and resolve urgent problems in local areas; and
2. Address the inability to propose non-quantitative/non-engineering performance indicators, such as what kind of social problems should be improved, the strength of public participation, and a local industrial upgrade or adjustment.

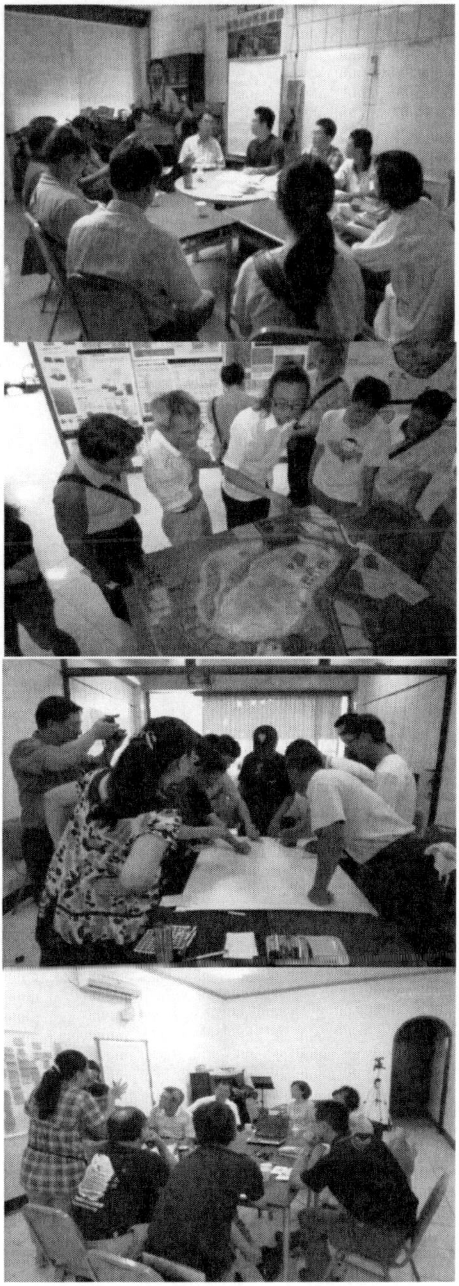

Figure 11. Workshops (© TIIWE).

Since 2006 up to 2012, 400 dialogue activities with a total length of 636 hours and 1,827 participants were held including 275 interviews, 86 group discussions, 32 workshops, and 7 conferences.

CONCLUSION: ONGOING EFFORTS SINCE 2012

A holistic plan is needed to resolve the land subsidence problem in Dongshi Township. As it involves comprehensive planning, three-stage public participation was organized to form local consensus and integrate the policies of related government agencies. In response, the central government has proposed the "Trans-Divisional Coastal Land Subsidence Improvement Plan for Dongshi of Chiayi County". The preliminary financial and implementation feasibility has been confirmed, thus laying a foundation for success in the future.

At present, the government has drawn up two-stage development strategies and physical measures with regard to the pilot study, land use, flood relief for low-lying areas, industrial adjustment, infrastructure renovation, and project management. Policy approval is in progress, and a sum of NT$1. 64 billion will be invested in the project over two phases: Phase I require about NT$806 million and Phase II needs about NT$835 million. It is projected that the investment will be totally returned from land development, and 1500 new working opportunities can be created, and the local economy and social conditions will be improved through the introduction of emerging industries in the project area.

FUTURE CHALLENGES

The bottom-up public participation which clearly identifies local development problems and local citizen needs and reaches a consensus on national land restoration with all stakeholder groups can help the government to effectively plan and implement policies. In the future, this method can benefit the national land restoration of the land subsidence areas in Dongshi Township, Chiayi County, and revitalize local economic development.

As professional facilitators, the authors attempt to help locals to reach a consensus to resolve problems through a bottom-up public participation process. However, as land subsidence is a complex issue, the national land restoration project in the area may face the following challenges:

1. The process of public participation and empowerment is quite slow. The public participation started in 2006 does not mean that local communities have been adequately empowered to resist the top-down policy-making process, or to monitor whether the government has faithfully implemented the consensus reached.
2. There are limitations to public participation, as the procedure and method of public participation are unorganized, and the public is not delegated with substantial decision-making authority. Therefore, the consensus reached may be changeable. In the future, in an attempt to influence the allocation of the government budget, stakeholder groups with substantial capacity, particularly the public sector and opinion representatives, may intervene in the promotion and implementation of the project during the process of policy implementation.
3. Bottom-up public participation is a dynamic interaction. The consensus reached is merely preliminary. In the future, the government will be challenged and questioned by all stakeholder groups when promoting any project in that area. Therefore, it is necessary for every project to reach an updating consensus through the public participation.
4. There are conflicts of interest among stakeholder groups. Therefore, it is still risky that the public interest and minority or underrepresented groups might be sacrificed in the name of project implementation in the future.

REFERENCES

Green Energy and Environment Research Laboratories, Industrial Technology Research Institute (2001) Analysis, investigation and monitoring in Taiwan subjected to Land subsidence, Water Resources Planning Institute, *Water Resources Agency of MOEA.*

Land Subsidence Database of Taiwan. (Available online: http://www. *subsidence.org.tw*)

Public Construction Commission, Executive Yuan (1998) *The Typhoon Herb disasters and reconstruction engineering documentary.*

Statistics of the Accounting and Statistics Department of Chiayi County Government. (Available online: http://www1.cyhg.gov.tw/accounting */chinese/CP.aspx?s=667&n=10967)*

Taiwan International Institute for Water Education (1997) Master Planning for Environmental Restoration and Sustainable Revitalization in Chiayi Coast Region, Water Resources Planning Institute, *Water Resources Agency of MOEA.*

Taiwan International Institute for Water Education (2011) Establishment and Assessment of Strategies for Public Participation and Empowerment in Coastal Subsided Restoration Program, Water Resources Planning Institute, *Water Resources Agency of MOEA.*

Taiwan International Institute for Water Education (2012) Fundamental Alignment of the standards of Process for Public Participation within the Planning for Water Resources Program, Water Resources Planning Institute, *Water Resources Agency of MOEA.*

In: Tsunamis
Editor: Tianxing Cai

ISBN: 978-1-62808-682-9

Chapter 3

Mixing and Sediment Processes Induced by Tsunamis Propagating Upriver

***Hubert Chanson*[1*1,2] *and Pierre Lubin*[†1]**
[1]Université de Bordeaux, Pessac, France
[2]The University of Queensland, Brisbane QLD, Australia

Abstract

The 26 December 2004 and 10 March 2011 tsunami disasters highlighted the rapid advance of the tsunami waters following rivers and canals causing massive damage deep inland. The leading edge of the tsunami formed a series of waves propagating upriver. This phenomenon is called 'shio-tsunami' (bore), 'kaisho' (tsunami-induced bore) or simply 'tsunami' (tidal wave) in Japan. After breaking, a tsunami wave propagating in shallow-waters is led by a bore. In these rivers and shallow-water bays, the propagation of the bores is associated with strong mixing and massive sedimentary processes upriver. This contribution reviews a number of tsunami-induced bore disasters before studying the tsunami propagation upriver. The aim of the contribution is to provide compelling evidences on the turbulent mixing and sediment processes

[*] The University of Queensland, School of Civil Engineering, Brisbane QLD 4072, Australia Email: h.chanson@uq.edu.au.
[†] Université de Bordeaux, I2M, Laboratoire TREFLE, 16 avenue Pey-Berland, Pessac, France, CNRS UMR 5295, Pessac, France E-mail: p.lubin@i2m.u-bordeaux1.fr.

induced by tsunami bores based upon some recent field observations as well as physical and numerical studies of tidal bores. It is shown that the propagation of the bore front modifies drastically the water column properties and the tsunami waters behind the front may carry a large amount of sediments, explaining the sedimentological and morphological changes observed after some recent disasters.

INTRODUCTION

Coastlines and rivers are places where large population densities with concentrated economical activities are found. It is critical to defend these regions against natural disasters including tsunami, one of the deadliest and most destructive natural hazards (Table 1). Rivers are natural breaches in coastlines, giving a natural path for the tsunami waters to flood inland areas. River mouths are often more vulnerable to tsunamis than other coastal areas protected by natural dunes, rock promontories and coastal structures. They are difficult to protect, but their critical funneling role is acknowledged in the extension of the inundations, sometimes further aided by the river discharge. This issue gained attention in the recent years, as tsunami waves shoaling into coastal regions are responsible for many phenomena such as inundations, wave propagation into rivers, erosion and overwash in sand spits (Shuto 1985, Tanaka et al. 2012a,b, Tinh et al. 2012).

A tsunami is a disturbance of the ocean surface usually caused by an earthquake triggering a displacement of the sea-bed, an underwater volcanic eruption or a landslide, either initiated underwater or sub-aerial (Abadie et al. 2012). A tsunami is a long-period wave with a wave length as large as 200 to 350 km. As the tsunami approaches a shoreline, its wave slows down and the wave height increases until it breaks. Major tsunami disasters were associated with well in excess of 430,000 losses of life (Table 1). Recent media documentaries showed the explosive impact of tsunami waves during the 2004 Indian Ocean tsunami and 2011 Tohoku tsunami, in a manner somehow similar to Hokusai's Great wave off Kanagawa (Hokusai 1826-1833) (Fig.1). Figure 1 shows a comparison between Hokusai's print and a typical tsunami warning sign post along a road to the Enshu coastline, Japan (Chanson 2010a). (The Enshu shoreline was historically affected by several tsunamis.) It must be acknowledged that Hokusai's Great wave off Kanagawa was likely a freak wave rather than a tsunami wave (Cartwright and Nakamura 2009).

The 26 December 2004 and 10 March 2011 tsunami disasters highlighted the rapid advances of the tsunami waters following rivers and canals causing massive inland damage (Fig. 2 & 3). The leading edge of the tsunami formed a series of waves propagating upriver. This phenomenon is called 'shio-tsunami' (bore), 'kaisho' (tsunami-induced bore) or simply 'tsunami' (tidal wave) in Japan. After breaking, a tsunami wave propagating in shallow-waters is led by a bore. In shallow bay systems, the process is somehow similar to a storm surge induced bore like in the Bay of Bengal (Sommer and Mosley 1972, Delwar 2001) or a tidal bore in some rivers (Chanson 2011a). The tsunami-induced bore may propagate far upstream. Some tsunami-induced river bores were observed in Hawaii in 1946 and in Japan in 1983, 2001 and 2003(Shuto 1985, Oshiki et al. 2008). During the 26 December 2004 tsunami catastrophe,tsunami-induced bores were observed in shallow-water bays and river mouths in Malaysia, Thailand and Sri Lanka (Fig. 2). Numerous evidences were documented during the 10 March 2011 tsunami in Japan as well as in the west coast of North America (Fig. 2). In these rivers and shallow-water bays, the propagation of the bores is associated with strong mixing and massive sedimentary processes upriver (Fig. 3).

When a tsunami wave propagates in a river, the wave characteristics are substantially modified. A surge is formed, a sudden water level increase is observed and some dispersive wave train occurs at the wave front (Tsuji et al. 1991, Yasuda et al. 2004, Tanaka et al. 2008). Bottom friction, river bed slope, width, water depth, upstream river discharge can affect the tsunami surge propagation. To date the physics of tsunami waves in rivers is poorly known. Yasuda (2010) conducted a numerical analysis of various wave models about tsunami ascending into rivers. The results suggested that, after careful calibration, computed and measured values agreed relatively well in terms of arrival time and maximum water level, but more complex phenomena such as breaking criterion of undular bore travelling against nonuniform flow have yet to be taken into account. There is a critical need for comprehensive studies of bore propagation in rivers to gain in forecasting capacity. Some recent works aimed at studying possible scenarios of tsunami propagation in the Columbia River (Yeh et al. 2012), the Targus River (Baptista et al. 2011) and the Suez Canal (Finkl et al. 2012). Numerical models were used to predict the inundation extents, but the limitations of the models were highlighted by the highly complex features of the flow into account: e.g., breaking of the bores, diffraction and wave dispersion, interactions with the river banks, energy dissipation due to friction or interaction with vegetation, soliton fission. Clearly it is critical to identify key physical factors that are important to tsunami hydrodynamics penetrating in rivers.

This contribution reviews a number of tsunami-induced bore disasters before studying the tsunami propagation upriver. The aim of the manuscript is to provide compelling evidences on the turbulent mixing and sediment processes induced by tsunami bores based upon some recent field observations as well as physical and numerical studies of tidal bores. It is shown that the propagation of the bore front modifies drastically the water column properties and the tsunami waters behind the front may carry a large amount of sediments, explaining the sedimentological morphological changes observed after some recent disasters.

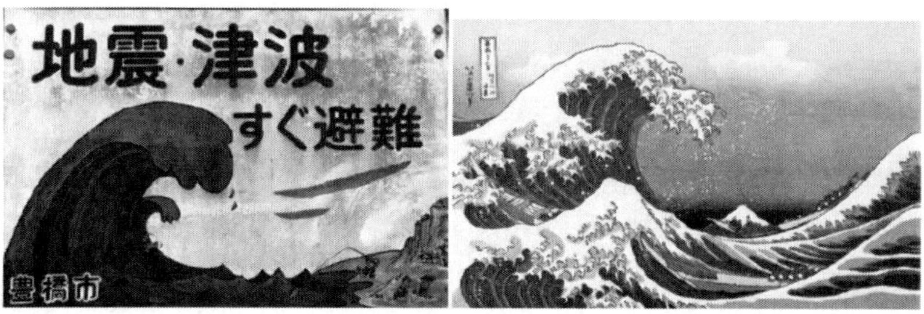

Figure 1. Tsunami warning sign off the Enshu coast (Japan) and HOKUSAI's Great wave of Kanagawa.

Figure 2-1. Photographs of tsunami-induced bores. (A) Tsunami-induced bore on 26 December 2004 at Penang (Malaysia) during the second series of tsunami waves (Courtesy of Angela EGOLD & Sonja PREIN) - Note the breaking bore in the foreground while the bore was undular in deeper waters (background).

Figure 2-2. Photographs of tsunami-induced bores. (B) Aerial shot of the tsunami-induced bore propagating upriver in the Naka River at Hitachinaka City, Ibaraki Prefecture (Japan) on 11 March 2011 - Bore propagation from right to left - The bridge is the Mito by-pass bridge (Road [6]) located about 11 km upstream of the river mouth.

Figure 2-3. Photographs of tsunami-induced bores. (C) Tsunami-induced bore propagating upriver in Emeryville, California (USA), on 11 March 2011 - The bore was induced by the Tohoku earthquake and tsunami in Japan.

Figure 3-1. Damage caused by tsunami-induced bore propagating upriver during the 10 March 2011 Tohoku earthquake and tsunami (Japan). (A) Aerial view the central part of the town of Minamisanriku along the Hachiman River, Miyagi Prefecture on 12 March 2011 - Inland damage were observed more than 3,000 m from the shoreline along the river path.

Figure 3-2. Damage caused by tsunami-induced bore propagating upriver during the 10 March 2011 Tohoku earthquake and tsunami (Japan). (B) Bore flow on the Hei River overtopping the left banks at Miyako, Iwate Prefecture, Japan on 11 March 2011 (Left) and 17 February 2012 (Right) - Close to the river mouth, looking downstream towards the river mouth (Courtesy of The Boston Globe, 7 March 2012).

Figure 3-3. Damage caused by tsunami-induced bore propagating upriver during the 10 March 2011Tohoku earthquake and tsunami (Japan). (C) Flooding induced by the bore in Miyako, Iwate Prefecture, Japan on 11 March 2011 (Left) and 17 February 2012 (Right) - Further upstream along the Hei River, looking downstream (Courtesy of The Boston Globe, 7 March 2012).

Table 1. List of some major tsunami disasters and associated losses of life

Date	Name	Losses of life
BC 1,500	Santorini island, Greece	--
30 July 1637	Gargano, Italy	5,000
27 January 1703	Shizoku Island, Japan	100,000
1 November 1755	Lisbon, Portugal	60,000
17 August 1883	Krakatoa, Indonesia	36,000
15 June 1896	Sanriku coast, Honshu, Japan	26,000
9 March 1957	Aleutian Tsunami	--
23 May 1960	Chile Earthquake & Tsunami	1,750
12 July 1993	Okushiri island, Japan	120
17 July 1998	Papua New Guinea	3,000
26 December 2004	Aceh, Indonesia	280,000
10 March 2011	Tohoku Earthquake & Tsunami	19,000

Note: (--): data not available.

FIELD OBSERVATIONS

The upstream propagation of tsunami-induced bores was documented in several scientific studies. Historical documents reported some flooding of the downtown of Lisbon (Portugal) by the rising of the waters of the Targus River during the 1755 tsunami (Baptista et al. 1998, 2011).

During the 26 May 1983 tsunami in the Sea of Japan, "*bores were observed in the Yoneshiro and Omono Rivers*" and "*the tsunami can be traced for up to 6.6 km from the mouth*" in the Omono River (Abe 1986); along a large river the tsunami travelled inland for about 15 km (Abe 1986). During the 2003 Tokachi-oki Earthquake and Tsunami, a tsunami propagated up several rivers on the Pacific coast of Hokkaido: "*six tsunami waves with a period of 30–40 min arrived at the mouth of the Tokachi River*"; "*these waves propagated up about 7 km from Ohtsu to Tabikorai in about 20 min*" (Yasuda 2010). Following the 27 February 2010 Chilean earthquake, tsunamiinduced bores were observed in several Japanese rivers, and "*the tsunami wave can be affected up to about 33 km upstream*" (Tanaka et al. 2011). A recent data analysis suggested that the inland propagation distance (from the river mouth) is inversely proportional to the river bed slope So (Kayane et al. 2012). About 500 people died from the 2010 Chilean tsunami. Many evidences of upriver tsunami propagation in Chile were detailed: large fishing boats were transported more than 10 km up the Maule River; video movies documenting tsunami propagation at least 15 km from the river mouth. Inundation and damage occurred more than a kilometer inland along several other rivers including at Coliumo and Tubul (Fritz et al. 2011, Koshimura et al. 2011).

In 2004, the Sri Lanka coastlines were hardly hit by the 2004 Indian Ocean tsunami and 35,000 people died during the event. The western coast of the island suffered substantial impacts. Thousands of plantations (rice, mango, banana, etc.) were almost entirely destroyed. Many boats vital to industrial and artisanal fisheries, a major economic activity, were washed away. Tsunami disasters in Sri Lanka were studied showing evidences of tsunami surges entering the rivers and travelling inland (Tanaka et al. 2008, Wijetunge et al. 2008, Wijetunge 2009a,b). In the Kalu River, a big wave was immediately followed by some smaller waves travelling up the river, propagating up to 20 km from the river mouth. The river path and its branching streams have had a significant influence on the distribution of inundation in Matara, a main city in southern Sri Lanka. In situ observations reported: "*rivers and streams that are open to the sea carry tsunami flood flow up stream in the form of a bore whilst spilling and inundating the low elevation terrain on either side*"; "*the surging tsunami front had first travelled over land and then fallen into water bodies such as lagoons, which in some instances, unable to absorb all of the overland flow, had spilt over and flooded inland areas that would otherwise not have received direct inundation*" (Wijetunge 2009a). The extent of inland inundation was thus facilitated by the river courses, streams, canals, lagoons and lakes. In other locations, the surging tsunami front moved across barrier spits, low points of dunes as well as sandbar formations at lagoon openings leading to flooding of onshore areas: "*the tsunami had climbed up the river and generated localized floods in the low-lying areas in the inlands*" (Tanaka et al. 2008).

The experiences of the 10 March 2011 Tohoku tsunami disaster highlighted tragically the turbulent transport induced by bores, including the transport of sediments, particulates and scalars with adverse implications in terms water quality and sediment processes. A survey data study summarised: "*the deepening of river water depth at the entrance due to tsunami impacts can cause a lot of problems such as floodplain erosion or deposition in the river due to waves further penetrating inside of river mouth*" (Tanaka et al. 2012a); "*many breaching places occurred along the coast and river mouth*", photography showed "*a huge morphologically changes*" of the river mouth: "*the river water depth and width are also become deeper and wider*" (Tinh et al. 2012). The Tohoku tsunami was observed to penetrate up to 50 km in the Kitakami River, resulting in widespread inundation along the river banks for about 15 km from the sea (Rao and Lin 2011). Observations indicated that on-land run-up decayed exponentially with increasing distance, up to 5 km, from the shoreline, but much longer tsunami run-ups were measured along rivers (Mori et al. 2011).

The environmental impacts of tsunami-induced bores encompass the destruction of farming communities, fishing industries, and touristic activities, as well as the destruction of crops, the seawater contamination of water supplies, surface waters and ground waters impacting on drinking water, irrigation and industrial water resources, but also the destruction of sewage lines and sewage treatment plants (Chanson 2005). Tsunami intrusion behaves differently when penetrating water bodies, compared to overland propagation (Adityawan et al. 2012). The intrusion distance was shown to be related to the bed slope, the river mouth type and the wave height after entering the river. Tsunamis are able to destroy most obstacles and structures at the river mouth, including constricted structures and sand spits (Tanaka et al. 2011, 2012b).

BASIC THEORY

Theoretical analysis of tsunami-induced bores is not straightforward because of the complexity of the processes involved when a tsunami wave penetrates a river channel (Caputo and Stepanyants 2003, Bukreev 2004, Pelinovsky and Rodin 2011). Tsunami wave shoaling and runup were shown to be facilitated and amplified in some bathymetric configurations (Didenkulova and Pelinovsky 2011). While the tsunami wave amplitude is moderate in deep waters, the tsunami slows down and the wave height increases near the shoreline. The wave height might reach several metres above the natural sea level. When a tsunami wave reaches a river mouth, its transformation into a bore and the upriver propagation of the bore may be predicted using the shallow water equations. For a one-dimensional unsteady flow, the equations of conservation of mass and conservation of momentum yield a series of two equations, called Saint-Venant equations:

$$\frac{\partial d}{\partial t} + \frac{A}{B} \times \frac{\partial V}{\partial x} + V \times \frac{\partial d}{\partial x} + \frac{V}{B} \times \left(\frac{\partial A}{\partial x} \right)_{d=\text{constant}} = 0 \tag{1}$$

$$\frac{\partial V}{\partial t} + V \times \frac{\partial V}{\partial x} + g \times \frac{\partial d}{\partial x} + g \times (S_f - S_o) = 0 \tag{2}$$

where d is the water depth, V is the flow velocity, t is the time, x is the longitudinal co-ordinate, A is the flow cross-section area, B is the free-surface width, and So is the bed slope and Sf the friction slope (Liggett 1994, Montes 1998). These equations cannot be solved analytically usually because of the non-linear terms that include the friction slope:

$$S_f = \frac{f}{2} \times \frac{V \times |V|}{g \times D_H} \tag{3}$$

where D_H is the hydraulic diameter, $|V|$ is the magnitude of the flow velocity and f is the Darcy-Weisbach friction factor that is a non linear function of Reynolds number and relative roughness height. The friction slope has the same sign as the velocity. Equations (1) and (2) may be rewritten as:

$$2 \times \frac{\partial C}{\partial t} + 2 \times V \times \frac{\partial C}{\partial x} + C \times \frac{\partial V}{\partial x} = 0 \tag{4}$$

$$\frac{\partial V}{\partial t} + V \times \frac{\partial V}{\partial x} + 2 \times C \times \frac{\partial C}{\partial x} + g \times (S_f - S_o) = 0 \tag{5}$$

where C is the celerity of a small disturbance equal to $C = \sqrt{g \times d}$ (Henderson 1966, Ligget 1994). Combining Equations (4) and (5), the Saint-Venant equations may be transformed into a characteristic system of equations:

$$\frac{D}{Dt}(V + 2 \times C) = -g \times (S_f - S_o) \qquad \text{forward characteristic} \tag{6}$$

$$\frac{D}{Dt}(V - 2 \times C) = -g \times (S_f - S_o) \qquad \text{backward characteristic} \tag{7}$$

along the respective characteristic trajectories :

$$\frac{dx}{dt} = V + C \qquad \text{forward characteristic} \tag{8}$$

$$\frac{dx}{dt} = V - C \qquad \text{backward characteristic} \tag{9}$$

where D/Dt is the absolute differential. For an observer travelling along the forward characteristics (Eq. (8)), Equation (6) is valid at any point; for an observer travelling on the backward characteristics (Eq. (9)), Equation (7) is satisfied everywhere.

With this transformation, the shallow water equations may be solved graphically. In the (x, t) characteristic diagram, the horizontal axis is the longitudinal distance x and the vertical axis is the time t (Fig. 4). Once the flow properties at t = 0 are known everywhere and the time-variations of the water depth at the river mouth are known, the upriver propagation of the tsunami wave and bore may be predicted. With the arrival of the tsunami wave, the water level at the river mouth increases rapidly with time. The backward characteristics issuing from the river mouth form a network of converging lines (Fig. 4). When two backward characteristics intersect, the water depth has two values at the same time: that is, the bore front. The bore forms at the first intersecting point. Afterwards, there is a water depth discontinuity along the forward characteristics forming the envelope of the bore front. The flow conditions across the bore leading edge satisfy the continuity and momentum principles.

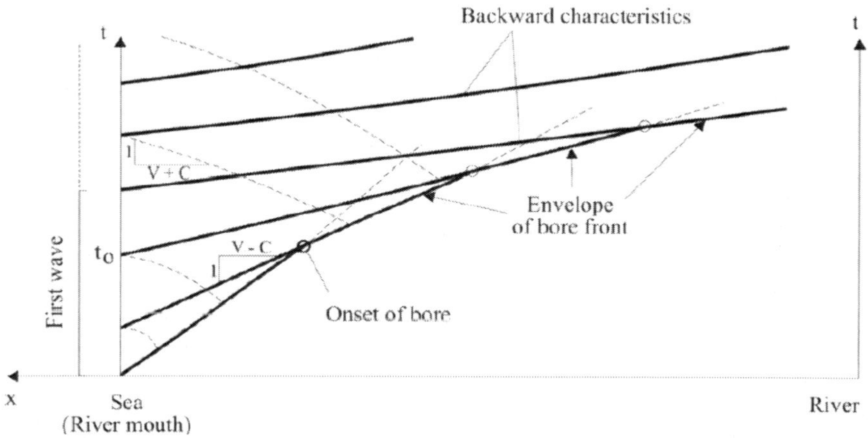

Figure 4. Characteristic diagram of a tsunami-induced bore propagating upriver.

The bore is a hydrodynamic shock and flow discontinuity (Lighthill 1978, Ligget 1994).In a system of reference following the bore front, the integral form of the continuity and momentum equations gives a series of relationships between the flow properties in front of and behind the bore (Rayleigh 1914, Lighthill 1978):

$$(V_1 + U) \times A_1 = (V_2 + U) \times A_2 \tag{10}$$

$$\rho \times (V_1 + U) \times A_1 \times (\beta_1 \times (V_1 + U) - \beta_2 \times (V_2 + U)) = \iint_{A_2} P \times dA - \iint_{A_1} P \times dA + F_{fric} - W \times \sin\theta \tag{11}$$

where V is the flow velocity positive downstream towards the river mouth, U is the bore celerity positive upriver (Fig. 5), ρ is the water density, A is the channel cross-sectional area measured perpendicular to the main flow direction, a momentum correction coefficient, P is the pressure, the subscript 1 refers to the initial flow conditions and the subscript 2 refers to the flow conditions immediately after the bore, F_{fric} is the flow resistance force, W is the weight force and the angle between the bed slope and horizontal. Neglecting the flow resistance, the effect of the velocity distribution and for a flat channel ($\theta \approx 0$), Equations (10) and (11) give a relationship between the ratio of conjugate cross-section areas A_2/A_1 as a function of the Froude number Fr_1 (Chanson 2012):

$$\frac{A_2}{A_1} = \frac{1}{2} \times \frac{\sqrt{\left(2 - \frac{B'}{B}\right)^2 + 8 \frac{B'/B}{B_1/B} \times Fr_1^2} - \left(2 - \frac{B'}{B}\right)}{\frac{B'}{B}} \qquad (12)$$

where the Froude number is defined as:

$$Fr_1 = \frac{U + V_1}{\sqrt{g \times \frac{A_1}{B_1}}} \qquad (13)$$

and B and B' are characteristic the free-surface widths defined as:

$$B = \frac{A_2 - A_1}{d_2 - d_1} \qquad (14)$$

$$B' = \frac{\iint_{A_1}^{A_2} \rho \times g \times (d_2 - y) \times dA}{\frac{1}{2} \times \rho \times g \times (d_2 - d_1)^2} \qquad (15)$$

Equation (12) is valid for any bore and hydraulic jump in an irregular channel (Chanson 2012). Its application was tested against some field data in tidal river bores (Fig. 6). Figure 6 illustrates the good agreement between Equation (12) and the field data. The effects of the channel crosssection irregularity increase with increasing Froude number and bore height (d_2-d_1).

For completeness, the solution of the momentum equation for a smooth rectangular channel is shown in Figure 6:

$$\frac{A_2}{A_1} = \frac{1}{2} \times \left(\sqrt{1 + 8 \times Fr_1^{\ 2}} - 1 \right) \tag{16}$$

The results highlight the limitations of the simplistic Bélanger equation (Eq. (16)) in natural irregular channels for which the cross-sectional properties do have a significant impact on the definition of the bore Froude number (Fig. 6).

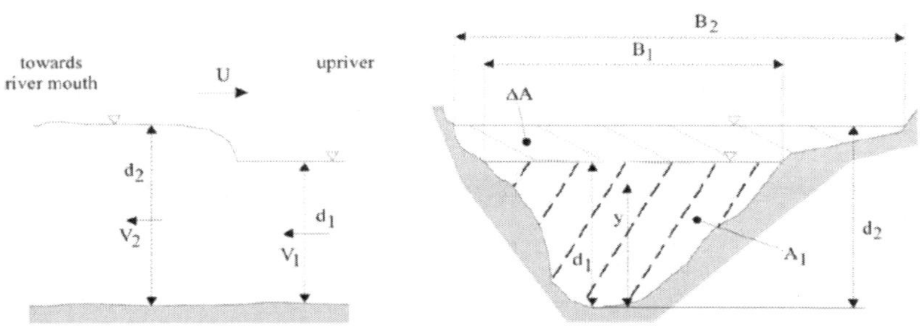

Figure 5. Definition sketch of a tsunami-induced bore propagating upriver.

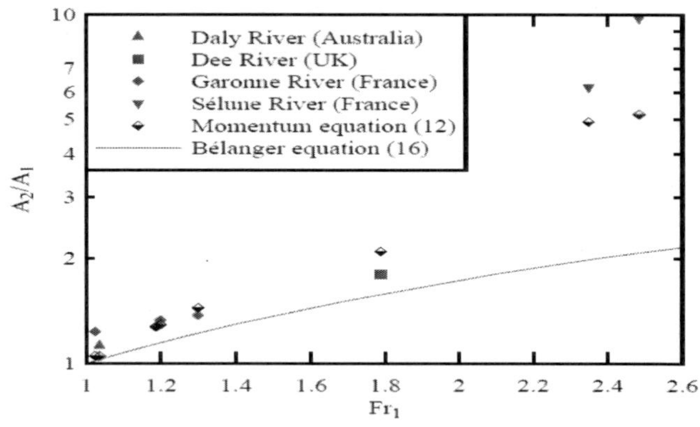

Figure 6. Semi-logarithmic relationship between the conjugate cross-sectional area ratio A_2/A_1 and Froude number Fr_1 - Comparison between the momentum equation (12), the Bélanger equation (Eq. (16)) for rectangular channels and tidal river bore data (Simpson et al. 2004, Wolanski et al. 2004, Mouaze et al. 2010, Chanson et al. 2011, Reungoat et al. 2012).

APPLICATION

The characteristic equations may be simplified when $S_f - S_o = 0$ anywhere anytime. A further simplification is the simple wave approximation defined as a wave for which $S_f = S_o = 0$ with initially constant water depth and flow velocity in the entire river section. Some detailed analytical solutions may be derived for the propagation of a shock in shallow waters (Poincaré 1910, Lacombe 1952, Chanson 2004).

Let us consider a river stretch, 3 km long and about 30 m wide, with a bed slope corresponding to a 1 m river bed drop over the reach. At the river mouth, an incoming tsunami wave imposes a time-variation of the water elevation as illustrated in Figure 7A. Initially the water depth was 2 m in the river and the mean velocity was 0.5 m/s.

Preliminary calculations are conducted assuming a simple wave solution for $0 < x < 3$ km and neglecting reflection effects. The results show formation of a first bore about 500 m upstream of the river mouth at t = 130 s. The bore propagates upstream and reaches the channel upstream end at t = 9 min when the bore height (d_2-d_1) is about 1.3 m. A second bore, associated with the sharp peak of the tsunami wave, forms at t = 12 min. The calculations yield further the water elevation and cross-sectional averaged velocity in the river (Fig. 7B). Figure 7B illustrates the longitudinal free-surface profile at t = 5 min (300 s).

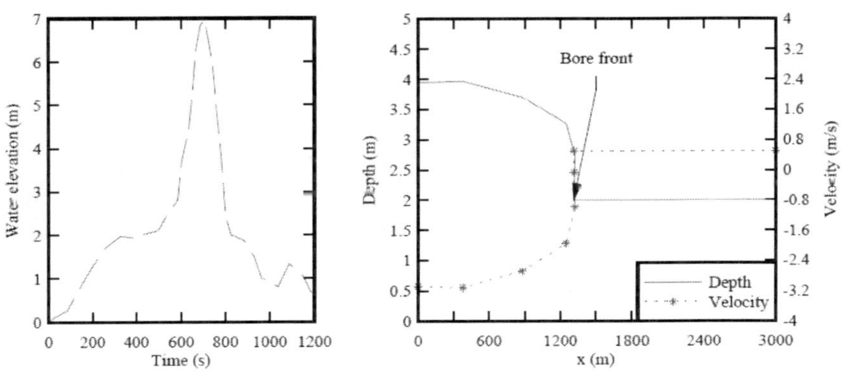

(A) Water elevation at the river mouth (B) Free-surface and velocity profiles at t = 5 min

Figure 7. Propagation of a tsunami wave into an idealised estuary.

TURBULENT MIXING IN BORES PROPAGATING UPRIVER: THE EXPERIENCE WITH TIDAL BORES

A tidal bore is an intense and powerful natural phenomenon observed in macro-tidal estuaries (Chanson 2011a). It generally occurs during spring tide conditions when the flood tide is confined to river mouths exhibiting converging funneled channel forms. When the tidal flow turns to rising, a positive surge propagates upstream the river to form the tidal bore. The bore development is closely linked with the tidal range and the river mouth shape, and its existence is sensitive to any small change in boundary conditions. The flow of the river often flows upstream after its passage (Chanson 2009, 2011c).

The existence of tidal bores is based upon an unstable hydrodynamic balance, which may be easily disturbed by changes in boundary conditions and freshwater inflow conditions. Despite numerous anecdotical evidences and archival documents, the tidal bore phenomenon remains poorly understood, rarely studied in details, due to its unsteady and turbulent nature. The number of field surveys and laboratory experiments of tidal bores has recently increased from the early historical descriptions (Chanson 2011a). Few numerical studies were conducted, based on the shallow-water equations or the Navier-Stokes equations, compared to experimental results. Numerical studies are restricted because of the complexity of the unsteady turbulent multiphase flow (Bombardelli and Chanson 2009), and the limited field data sets available for validation. Tidal bores also imply free-surface deformations, a large variety of length scales and air entrainment, which cannot be taken into account by many numerical tools. Madsen et al. (2005) developed numerical schemes dedicated to the nonlinear shallow water equations to allow for the treatment of tidal bores in the Qiantang River (China). Comparative results were shown, but an extensive calibration phase involving a sensitivity study of boundary conditions, bottom friction, and local bathymetric modifications was necessary. Unfortunately no velocity measurement was available and more information about the net currents in the Qiantang River estuarine zone is needed. A more recent attempt was based upon the Navier-Stokes equations, and the results were compared to laboratory experiments (Furuyama and Chanson 2010). Some interesting flow features were observed, although the results could be improved with a finer mesh grid resolution and more accurate numerical schemes, as discussed by Lubin et al. (2010 a,b). The CFD modelling based upon the (full) Navier-Stokes equations give access to the most intricate flow features (aeration of the flow, unsteady turbulent flow motion, coherent structures), assuming that a sufficient number of grid points are used for discretisation.

The bore is a series of waves propagating upstream as the tidal flow turns to rising. Two different types of tidal bores are generally observed depending upon the Froude number Fr_1 (Chanson 2009). For Fr_1 less than 1.4 to 1.6, tidal bores develop as undular bores characterised by a leading front followed by a train of advancing undulations, called whelps, and an absence of wave breaking, but close to the banks (Fig. 8A). Breaking bores are rarer ($Fr_1 > 1.6$), often restricted to high tide conditions and localised in some estuarine sections (Fig. 8B). About 95 % of the observed tidal bores are undular bores (Chanson 2011a). Experimental results indicated systematically some basic flow features. The tidal bore is characterised by a sudden increase in water depth, yielding a rapid change in longitudinal velocity to satisfy the conservation of mass. In a natural river, a flow reversal is often observed. The longitudinal velocities highlight a rapid flow deceleration at all vertical elevations during the bore front passage, while some large fluctuations of longitudinal, transverse and vertical velocity components are observed beneath and behind the tidal bore. The tidal bore is basically a hydrodynamic shock followed by a highly turbulent flow motion with significant fluctuations of all velocity components.

Figure 8A. Photographs of undular and breaking bores. (A) Undular tidal bore of the Colorado River (Mexico) on 13 October 1985 (Courtesy of Steve Nelson) - Bore propagation from right to left.

In an undular tidal bore, the longitudinal velocity oscillates with time with the same period as, but out of phase with, the free-surface undulations immediately after the front (Koch and Chanson 2008, Chanson 2010b, Chanson and Tan 2010). The undular bore front passage is associated with a relatively gentle longitudinal flow deceleration observed at all vertical elevations. The longitudinal velocity component is found to be minimum beneath the first wave crest and it oscillates afterwards with the same period as the surface undulations and out of phase. The vertical velocity presents a similar oscillating pattern beneath the free-surface undulations with the same periodicity, and out of phase. Maximum velocities are observed beneath the wave troughs and minimum velocities below the wave crests. This trend is seen at all vertical and transverse locations. The observations were reported in a number of physical observations.

Figure 8B. Photographs of undular and breaking bores. (B) Breaking tidal bore of the Sélune River (France) on 24 September 2010 (Courtesy of Dominique Mouaze) - Bore propagation from right to left.

The breaking tidal bore data show a sharp front with a marked roller and some air entrainment (Fig. 8B). The free-surface is sometimes curved upwards immediately prior to the roller toe. The gentle rise of the free-surface may be linked with a gradual decrease of the longitudinal velocity component at all vertical elevations (Hornung et al. 1995, Koch and Chanson 2009, Docherty and Chanson 2012). The roller passage is associated with a sudden decrease of the longitudinal velocity component. The flow deceleration is notably sharper than that for an undular bore at all vertical elevations. Close to the bed, the instantaneous longitudinal velocity can become negative, highlighting a relatively rapid transient associated with some unsteady flow separation (Koch and Chanson 2009, Chanson 2010b, Mouaze et al. 2010, Docherty and Chanson 2012). The velocity measurements highlighted some energetic turbulent events beneath and behind the tidal bore front. Some sudden and rapid fluctuations of the transverse and vertical velocity components are experimentally observed (Koch and Chanson 2009, Chanson 2010b), while some CFD modelling hinted the production of large turbulent eddies beneath the bore front and their upstream advection behind the bore (Lubin et al. 2010a,b). These vortical structures remain next to the bed as the bore propagates upstream, while the presence of persisting coherent structures indicates that a great amount of sediment matters could be placed into suspension and advected upstream in a natural estuary.

The visual observations of tidal bores highlight the turbulent nature of the propagating bores (Fig. 8) (Chanson 2011a). Some energetic turbulent events are observed in natural rivers, probably due to some form of macro-turbulence likely induced by some secondary motion. Kjerfve and Ferreira (1993) also account for the violence of the flows in the Rio Mearim (Brazil): "*Still, the field installations posed great difficulties. On 30 January 1991, one sawhorse and instrument tumbled along the bottom for 1.4 km with currents exceeding 3 m/s, was buried in a sand bank, and had to be abandoned although the instrument was recovered.*" In the River Dee (UK), Simpson et al. (2004) reported similar problems: "*During this second deployment, the instrument was repeatedly buried in sediment after the first tidal cycle and had to be dug out of the sediment, with considerable difficulty, at the time of recovery.*" In the Daly River (Australia), Wolanski et al. (2004) described a period of very strong turbulence observed about twenty minutes after the bore passage, lasting for about three minutes: "*This unsteady motion was sufficiently energetic to topple moorings that had survived much higher, quasi-steady currents of 1. 8 m/s (Wolanski et al., 2001)*" (Fig. 9B). In the Seine River (France), some field observations were undertaken in 1855 (Bazin 1865): "*Floats were introduced at several heights prior to the arrival of the tidal bore. The float direction was observed after the bore passage. In one case, the surface float continued to float downstream for about 2 min after the bore, while the bottom float flowed downstream only for 1.5 min. Afterwards the floats flowed upstream.*" Repeated field work accidents were reported in the literature. Chanson et al. (2011) performed a detailed study of turbulent and sedimentary processes in the tidal bore in the Garonne river (France) at the end of a dry summer, while Reungoat et al. (2012) surveyed the same site, but a few weeks after a major flood: "*During the field deployment, the authors experienced a major problem: the ADV stem was bent along the main upstream flow direction. The authors found the damaged unit when it was retrieved at the end of the study. [...] Based upon the visual observations and ADVrecord, it is believed that the ADV unit stem was hit by submerged debris during the early flood tide...*" Mouaze et al. (2010) conducted field measurements in the tidal bore of the Sélune River, in the Bay of Mont Saint Michel (France). They used an ADV system fixed to a tripod, consisting of three poles driven into the bed sediment materials and tightly fastened down together. "*About 40 s after the passage of the bore, the metallic frame started to move. Vibrations were followed by local liquefaction around each pile which then lost their friction stability inside the mobile bed. One pole was lost, and the ADV support failed completely 10 minutes after the tidal bore.*" [The next morning] "*the remaining poles were dug out of the bed sediments to avoid any damage to the ADV unit during the tidal bore passage. It took three people nearly 25 minutes to remove the poles in the darkness of the early morning.*" Reungoat et al. (2012) observed some unusual flow reversal delay between the bore front passage and longitudinal flow reversal. The recorded data showed the reversal in longitudinal flow direction about 50 s after the bore front, while the free-surface velocity next to the survey staff reversed direction about 6 s after the bore front. This unusual feature has been sometimes reported. In the Severn River (UK), some delayed flow reversal was seen depending upon the relative water elevation and bore strength: "*the water near the bed still flowing downwards for up to ten minutes after the surface has been suddenly reversed by the passage of a fairly large bore*", "*with small bores the normal downward flow comes gently to a standstill after the bore had gone by and it may be a minute before the upward stream gathers momentum*" (Rowbotham 1983). Kjerfve and Ferreira (1993) indicated some unusual flow reversal pattern in the Rio Mearim (Brazil): "*At times, the downstream flow resumed after passage of the bore for another 30 seconds before the flow again surged upstream.*" these observations were based upon measurements at 0.7 m above the bottom. Kjerfve and Ferreira (1993) reported further an early flow reversal: "*the current began changing directions 1 min ahead of arrival of the bore*". Further field studies highlighted some unusual mixing processes. Chanson (2011c) observed some transient intrusive fronts behind the tidal bore in Bay of Mont Saint Michel (France), inducing further secondary currents and vertical circulation. The front arrived typically a couple of minutes after the bore front and lasted several minutes.

These observations demonstrate unequivocally the complicated flow features and turbulent mixing associated with tidal bore propagation. Large energetic vortical structures are characteristic features of tidal bores. It is thought that these are induced by some secondary current motion and enhanced in the natural, three-dimensional, non-prismatic irregular and meandering channels (Fig. 9). The measurements of large transverse and vertical velocity fluctuations behind the tidal bore implied the existence of transient secondary currents behind the bore front, associated with some unsteady transverse shear pattern (Fig. 9A). Detailed measurements of the turbulent characteristics have been performed in natural rivers and experimental facilities. Wolanski et al. (2001) presented field studies on the sediment dynamics of the two arms of Cambridge Gulf, tropical Western Australia., highlighting the recent existence of tidal bores in this river. This phenomenon only appeared due to successive human interventions, leading to significant changes in the tidal dynamics and bathymetry of the estuary. Wolanski et al. (2004) presented field measurements undertaken in the Daly River (Australia) and observed a three-minute duration patch of macro-turbulence that occurred 20 minutes after the undular bore front (Fig. 9B). Simpson et al. (2004) studied tidal bores propagating in the tidally energetic estuarine channel of the Dee River (UK). They observed a high degree of variability in bore intensity, this unpredictable aspect being characteristic of many tidal bores in natural channels. The bore front involved an almost instantaneous injection of turbulent kinetic energy, resulting in large turbulent kinetic energy levels which rapidly dissipated. The large bottom shear stresses also generated large turbulent kinetic energy levels in the strong flood currents (1.5 m/s). For comparison, the ebb flow was found to be relatively tranquil with maximum speed of 0.5 m/s, thus inducing smaller turbulent kinetic energy production rates.

The Reynolds stresses characterise a transport effect resulting from turbulent motion induced by velocity fluctuations with its subsequent increase of momentum exchange and of mixing. Physical data generally showed large normal and tangential Reynolds stresses beneath the bore front. A comparison between undular and weak surge data suggested some basic difference (Koch and Chanson 2009). In breaking surges, large stresses are observed in regions of high velocity gradients, while some unsteady flow recirculation are recorded next to the bed. In undular surges, the largest Reynolds stresses are recorded in the lower flow region including next to the bed, and maximum normal and tangential stresses are observed beneath the wave crests behind the advancing front. Experimental measurements show large turbulent stresses and turbulent stress fluctuations, below the bore front and ensuing flow. Beneath the positive surge with roller, large normal and tangential Reynolds stresses were observed in the upper water column (Chanson 2010b, Chanson and Docherty 2012). The Reynolds stress magnitudes are significantly larger than before the surge passage at all vertical elevations. In the undular surge, the largest normal and tangential stresses are observed beneath the wave crests and just before each crest (Koch and Chanson 2008, Chanson 2010b).

Both Mouaze et al. (2010) in the Bay of Mont Saint Michel (France) and Chanson et al. (2011) in the Garonne River (France) found that the Reynolds stress measurements indicated some large and rapid fluctuations during the tidal bore and flood flow. The Reynolds stress magnitudes were significantly larger than during the ebb tide, and some substantial normal and tangential stress fluctuations were observed. This was confirmed by Reungoat et al. (2012) showing that the turbulent stress magnitudes were larger than during the ebb tide when large fluctuations in normal and tangential stresses were observed. The time-variations of shear stresses highlighted a drastic increase by nearly one order of magnitude in shear stress levels during the flow reversal itself.

Chanson (2011b) showed the effect of channel constriction and bridge piers on the propagation of undular bores. The data illustrated the major impact on the free-surface properties, as some strong macro-scale turbulence was produced when the tidal bore propagated through the constriction, and further vorticity was produced by and advected behind the bore front. This is consistent with in situ observations. The effects of bed roughness were shown by Chanson (2010b), Chanson and Docherty (2012) and Docherty and Chanson (2012). For an undular bore, the bed roughness induced a strong attenuation of the oscillating free-surface flow pattern effect on the longitudinal and vertical velocity components next to the bed, although this quasi-periodic pattern was still seen close to the free-surface. In the breaking bore, the longitudinal velocity data suggested a longer transient recirculation next to the invert on the rough bed. The height and duration of the transient were a function of the bed roughness, with a higher and longer recirculation region above the rough bed. The vertical velocity data presented some positive, upward motion during the bore passage, with increasing maximum vertical velocity with increasing distance from the bed. The transverse velocity data showed some large fluctuations after the roller passage, which highlighted some intense secondary motion advected behind the bore. Such a recirculation pattern is significant because it may contribute to very intense bed scouring beneath the tidal bore in presence of coarse bed materials and bed forms. Chanson et al. (2012) argued that all experimental studies conducted in laboratories are based upon a Froude similitude, but with Reynolds numbers about one to two orders of magnitude smaller than in most natural estuarine systems. Further all physical arrangements in laboratory did not reproduce the 'true' tidal bore motion observed in a natural estuary.

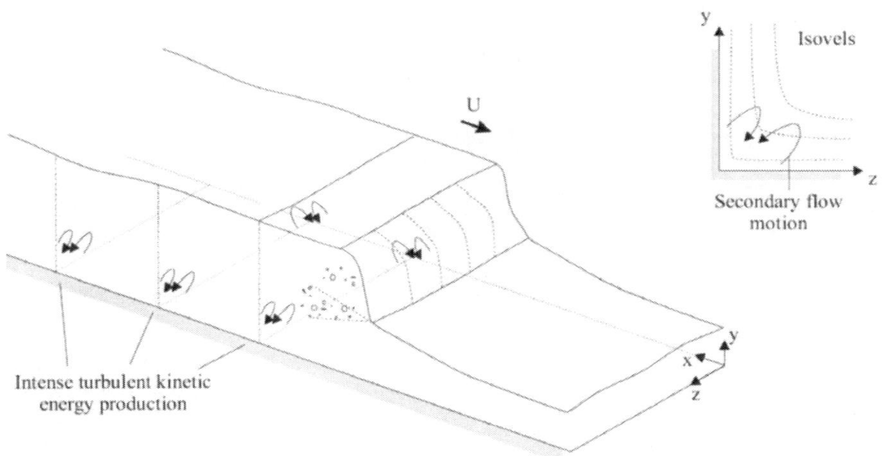

Figure 9A. Macro-turbulence generated during tidal bore propagation. (A) Near-bed turbulence and secondary currents in a tidal bore propagating in a rectangular channel - Right: secondary flow motion in the corner.

The structure of the flow field still has to be detailed in field surveys, and no visualisation technique was implemented to illustrate the coherent structures detected by some field and numerical studies. Scale effects may occur, but have still to be quantified. Some studies indicate that bed roughness affects the tidal bores, but systematic studies still have to show the effect on the structure of the flow. Moreover, the transition between undular and breaking tidal bores has to be investigated. Overall all the data sets present instantaneous turbulent stresses larger than the onset of sediment motion for cohesive and non-cohesive sediment materials, and the observations are consistent with the evidences of major sedimentary processes in tidal bore affected estuaries. Instantaneous Reynolds stress data highlight large turbulent stress magnitudes and turbulent stress fluctuations during the passage and behind the bore front at all vertical elevations. Hence bed erosion may take place during the surge front passage, and the eroded material and other scalars are advected in the "whelps" and wave motion behind the tidal bore front. These would be consistent with very-strong turbulent mixing observed in tidal bore affected estuaries, associated with accretion and deposition in the upper estuarine zones.

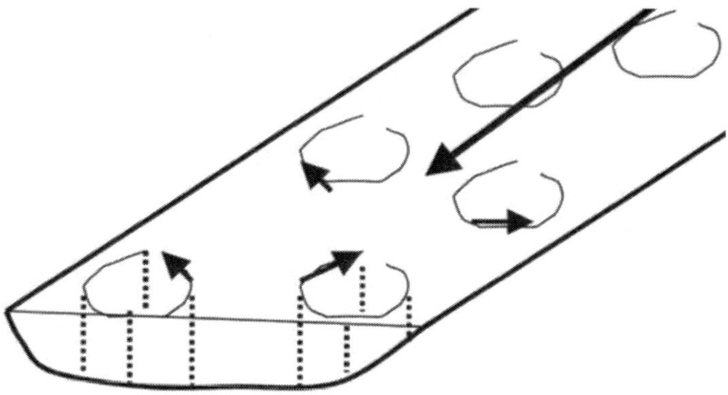

Figure 9B. Macro-turbulence generated during tidal bore propagation. (B) Surface macro-turbulence and scars advected upstream behind the undular tidal bore of the Daly River (Australia) and observed 20 min after the bore passage (after WOLANSKI et al. 2004).

SEDIMENTARY PROCESSES IN BORES PROPAGATING UPRIVER

A bore induces a strong turbulent mixing which can affect both estuarine and river eco-systems. Kjerfve and Ferreira (1993) reported that "*in shallow areas, the water boiled violently after passage of the bore and became brownish-black in color and heavily sediment-laden due to turbulent scouring and resuspension of bottom sand and mud.*" Tessier and Terwindt (1994) showed a major role played by tidal bores in the estuarine zone of the Bay of Mont Saint- Michel. Large sedimentary structures undergo large deformations during tidal bores passages. Chen (2003) analysed some surface water samplings under a 0.6 m-high breaking bore: the sediment concentrations taken in the bore front was up to 15.92 kg/m-3; two hours later the sediment concentration was down to 2.50 kg/m-3, illustrating the stronger turbulent bore capacity to carry a large amount of sediments. Wolanski et al. (2004) observed a time lag between the passage of an undular bore and resuspension, as accelerating flood tide currents resuspended the sediment two minutes after the bore front. For comparison, the resuspension of sediments under many tidal bores are generally quasi-instantaneous (Wolanski et al. 2001, SIMPSON et al. 2004, Mouaze et al. 2010, Chanson et al. 2011, Furgerot et al. 2012). The tidal bores do impact onto the estuarine eco-system (Donnelly and Chanson 2005). Tidal bores are turbulent flows, associated with strong mixing and bed material scouring. As the bore propagates, air is entrained in the roller of the breaking bore and the river flows upstream, advecting suspended matters. Donnelly and Chanson (2005) argued that undular tidal bores have some potential to scour the channel bed. Upward flow motion between troughs and subsequent crests advects bed material towards the free-surface. The rapid velocity redistributions, associated with the long-lasting chaotic wave motion, contribute to maintain sediment suspension which is advected upstream with the bore and eventually deposited in intertidal zones. A comparison between undular and weak surge data suggested two main mechanisms of sediment scour and scalar transport (Koch and Chanson 2009). In weak surge flows, the data indicate some rapid flow separation beneath the surge front. Flow separation and large transverse velocities enable scour and erosion of unconsolidated bed materials that are subsequently mixed in the developing shear region of the roller and advected by large coherent structures behind the roller. In undular surges, maximum Reynolds stresses are observed beneath and just before each wave crest. Bed erosion may take place beneath each wave crest, and the eroded material and other scalars are advected in the "whelps" and wave motion behind the first wave crest. The long lasting impact of the free-surface undulations is a key feature of undular tidal bores in natural systems (Chanson 2010b,c).

The macro-turbulent eddies observed in natural rivers "*are important in the transport of fine sediment because, [...] they essentially double the suspended sediment concentration*" (Wolanski et al. 2006). In the Sélune River, the passage of the bore and the ebb flow was characterised by large suspended sediment concentrations (Mouaze et al. 2010). In the Graonne River, some unusually high suspended sediment concentration about 100 s after the tidal bore front, this event lasting for more than 10 minutes (Chanson et al. 2011). It has been speculated that the tidal bore passage scoured the bed and convected upwards the bed material, reaching the free-surface after the bore passage: "*behind the tidal bore, the net sediment flux magnitude was 30 times larger than the ebb tide net flux and directed upstream. A striking feature of the data set was the intense mixing and suspended sediment motion during the tidal bore and following flood tide.*" Reungoat et al. (2012) observed that, "*the suspended sediment flux data indicated a downstream positive suspended sediment flux during the end of the ebb tide prior to the tidal bore. After the passage of the bore, the net sediment mass transfer per unit area was negative (i.e. upriver) and its magnitude was 1.5 to 2 times larger than the ebb tide net flux.*" In the Qiantang River (China), "*sediment concentration rapidly and substantially increased from roughly 0 kg.m^{-3} to greater than 70 kg.m^{-3} at the Ershigongduan station and rose from roughly 2 kg.m^{-3} to 20 kg.m^{-3} at the Ershiergongduan station*" (Pan and Huang 2010). Concerning the tidal bore influence in the Qiantang river (China), Fan et al. (2012) stated: "*the tidal bore and subsequent rapid flood flows take a leading role in shaping estuarine morphology and depositional strata, even though they only account for about one tenth of a single semidiurnal tidal cycle.*"

Chanson and Tan (2010, 2011) illustrated the turbulent mixing capacity of tidal bores and longitudinal dispersion of fish egg particles. The large turbulent structures present under tidal bores were responsible for the vertical water mixing as a tidal bore propagates upstream in an estuary. These large-scale eddies were also responsible for the rapid longitudinal dispersion of the fish eggs, also seen in the field (Rulifson and Tull 1999). Some form of preferential motion seemed to depend upon the particle's vertical elevation. A key result was the comparative observations suggesting that undular bores induced a larger particle mixing compared to breaking bores, especially in the upper flow region.

Khezri and Chanson (2012a,b) studied the inception of sediment motion and transient sediment transport beneath an unsteady tidal bore flow, both theoretically and physically. A key result is the identification of the dominant contribution of the longitudinal pressure gradient force to sediment inception beneath the breaking bore. The sediment particles are de-stabilized by the roller toe passage beneath the breaking bore front and advected upstream by bed load motion. A transient recirculation next to the bed leads to a delayed drag force contribution acting in the upstream direction and adding to further transport of the particles. The undular bore had a negligible effect on gravel particle movement, likely linked with the large sediment size (Khezri and Chanson 2012b). Under a breaking bore, the magnitude of negative transient longitudinal velocity close to the bed was greater on fixed bed than on mobile bed, indicating some damping effect of the movable gravel bed (Khezri and Chanson 2012c). Simply tidal bores do have a significant impact in terms of sediment processes on an estuarine system, and this was evidenced during tsunami-induced bores.

DISCUSSION

Tidal bores are sometimes presented as a proxy to tsunami-induced river bores, although some limitations must be clearly stated. A tidal bore is a predictable event induced by the tides, which are caused by the gravitational influences of the moon and sun. The tides and tidal waves are most pronounced in narrow bays or in rivers along the coast. On the other hand a tsunami is not related to tides: it occurs when some unpredictable, highly-energetic and rapid event disturbs the ocean, although the impact of a tsunami could be influenced by the tidal level when it reaches the shoreline. Tsunamis and tidal bores have different length and time scales. The time scale of a tidal bore is 12 h 25 min or 24 h 50 min for semi-diurnal and diurnal tidal regimes respectively, while the time scale of a tsunami-induced bore may vary from a few minutes to more than 1 hour. A tidal bore is simply a true tidal wave and it is not to be confused with a tsunami, which is a large ocean wave travelling primarily on the open ocean.

Some similarities between tidal bores and tsunamis propagating in rivers can also be highlighted. Both flow motions are similar, with massive volumes of water rushing upriver. Both types of bores are neither solitons nor progressive waves, but rather look like large walls of water that keeps coming from the sea. In both cases, there is a net flux of mass and momentum during the bore passage. Tidal bores and tsunamis are not solitary waves (Madsen et al. 2008), although solitary waves are known for having some interesting properties: e.g., symmetrical form with a single hump, uniform velocity without changing form. The tide does not consist of a single wave, and there is now clear evidence that tsunamis are not solitary waves (Madsen et al. 2008).

Both tidal bores and tsunamis originate from the ocean. The wave generated by a sudden and energetic event may travel for thousands of kilometres across the ocean until it reaches the coastlines. While it propagates in the ocean, the wave height is only the order of a metre or less, and the wave length can be more than several hundred kilometres, making it difficult to detect. In shallow waters, the energy is concentrated by shoaling and often tunnelling due the bathymetry, inducing some drastic wave steepening. Tides in the open ocean are also usually of much smaller amplitude than those along the coast, partly because of amplification by reflection and resonance. When approaching an inlet whose shape focuses the tidal wave energy, the funnelling effect induced an increase of tide height. More generally the tidal wave is shoaling into shallower water, its celerity decreases, and both tide height and tidal current strength increase accordingly. Basically both tidal bores and tsunamis are more spectacular when they arrive in funnel-shaped shallow water areas. Tidal bores can reach large heights (e.g. Amazon River), they are able to of overtop dikes and levees (e.g. Qiantang River) and can modify drastically the bed morphology (Chanson 2011a). In the Sélune River (France), Chanson (2011a) reported "*the formation of a new main channel by the tidal bore on 31 August 2008*"; "*The tidal bore cut a channel meander*"; "*The channel incision event was followed by intense bed form motion processes and standing waves during the early flood tide flow*" (Fig. 10). Figure 10 illustrates the leading edge of the bore cutting a new channel in the Bay of Mt St Michel (France).

Figure 10. Channel incision by the tidal bore of the Sélune River (France) on 31 August 2008 - The red arrow points to the bore front cutting a new channel - Bore propagation from background to foreground.

Both flows need some converging paths to propagate far upstream in the river. The mouth shape of the estuary is critical to facilitate the focusing of energy. If the shape is too narrow, both tidal waves and tsunamis will dissipate the energy by breaking. Both types of bore can propagate under the form of undular or breaking bores in the rivers (Tsuji et al. 1991). Many video footages and photographs of tsunami propagation upriver during the 11 March 2011 Tohoku tsunami showed the formation of an undular bore. The formation of undular bores was observed during the 1983 Nihonkai-Chubu tsunami by Shuto (1985), who reported that short waves of the order 10-15 s were riding on top of the main tsunami, which was of the order 5-30 minutes (also Madsen et al. 2008, Madsen 2010). Tidal bores can reach long upstream distances (hundreds of kilometers) with a typical celerity ranging between 15 to 20 km/h. Tsunami-induced bores were reported to travel upstream rivers as fast as tidal bores (Roh et al. 2012).

A tidal bore is a natural and fragile phenomenon, which is of great importance for the ecology of an estuary, whereas a tsunami is a catastrophic event. The tidal bore is a very important phenomenon for the estuarine ecosystem: e.g., turbulence, mixing capacity, sediment transport, remobilisation of particulates and pollutants. Tidal bore affected estuarine systems are known feeding and spawning grounds for various species. Carnivores and scavengers are common behind tidal bores. In the Amazon, piranhas gobble up fish and crabs. Crocodiles swim behind the Styx River bore in Queensland (Australia). The disappearance of several bores was reported to adversely impact onto the ecosystem, with elimination of fish species. Tidal bores also impact human industrial activities. For example, some sections of the Airbus A380 travel on barges on the Dee and Garonne River estuaries that are both affected by tidal bores. The impact on the ecology is acknowledged (Chanson 2009, 2011c). Tidal bores can tear vegetation like trees from their roots, while tidal bore affected estuarine systems are known feeding and spawning grounds for various species (Rulifson and Tull 1999). Major modifications of tidal bores can lead to fish species' disappearance. Tidal bores are tourist attractions in many locations. Numerous places are known for surfers' gatherings. Meetings and cultural event are organised during spring tide periods to coincide with the occurrence of large tidal bores. People are often regrouped in associations interested with the observation of this natural phenomenon, and numerous web sites share the experiences of worldwide tidal bores. The tidal bore has become some fragile heritage: people can watch television shows and often read articles in magazines and daily newspapers concerning tidal bores.

Man-made interventions led to the disappearance of several bores: e.g., harbour facilities, bridges, locks, dams, dredging operations. The global climate change is also responsible for the modifications of the conditions of appearance of bores: variations of river flows and bathymetries, rise of water and sea levels, floods, etc. The famous tidal bore of the Seine River no longer exists following dike constructions and dredging operations in the 1960s.

CONCLUSION

The 26 December 2004 and 10 March 2011 tsunami disasters highlighted the rapid advances of the tsunami waters following rivers and canals causing massive damage deep inland. The leading edge of the tsunami formed a series of waves propagating upriver. In these rivers and shallowwater bays, the propagation of the bores is associated with strong mixing and massive sedimentary processes upriver. A number of tsunami-induced bore disasters were reviewed herein and the basic equations of tsunami propagation upriver were developed. A number of compelling evidences on the turbulent mixing and sediment processes induced by tsunami bores were presented based upon some recent field observations, as well as physical and numerical studies of tidal bores. It was shown that the propagation of the bore front modifies drastically the water column properties and the tsunami waters behind the front may carry a large amount of sediments, explaining the sedimentological and morphological changes observed after some recent disasters.

Tidal bores and tsunamis propagating upriver are two distinct phenomena, showing some interesting similarities when propagating upriver. Human activities can modify or even prevent the occurrence of tidal bores. The same may be considered to minimise the impact of tsunamis propagating upriver.

ACKNOWLEDGMENTS

The manuscript was peer-reviewed by Professor Hitoshi TANAKA (Tohoku University, Japan) and Dr Hubert BRANGER (IRPHE/Luminy, France). The authors thank them of their helpful comments. They acknowledge the discussions with their students and co-workers. The financial assistance of the Agence Nationale de la Recherche (Projet MASCARET 10-BLAN-0911-01) is acknowledged.

REFERENCES

Abadie, S., Harris, J. C., Grilli, S. T., and Fabre, R. (2012). "Numerical Modeling of Tsunami Waves generated by the Flank Collapse of the Cumbre Vieja Volcano (La Palma, Canary Islands): Tsunami Source and

Near Field Effects." *Journal of Geophysical Research C: Oceans*, Vol. 117, No. 5, C05030.

Abe, K. (1986). "Tsunami Propagation in Rivers of the Japanese Islands." *Continental Shelf Research*, Vol. 5, No. 6, pp. 665-677.

Adityawan, M.B., Roh, M., Tanaka, H., and Farid, M. (2012). "The Effect of River Mouth Morphological Features on Tsunami Intrusion." *Proc. Annual International Conference organized by the International Institute for Infrastructure*, Renewal and Reconstruction (IIIRR), Kumamoto, Japan, August 24 - 26, pp. 75-83.

Baptista, M. A., Heitor, S., Miranda, J. M., Miranda, P., and Mendes Victor, L. (1998). "The 1755 Lisbon Tsunami; Evaluation of the Tsunami Parameters." *J. Geodynamics*, Vol. 25, No. 2, pp. 143-157.

Baptista, M. A., Miranda, J. M., Omira, R., and Antunes, C. (2011). "Potential Inundation of Lisbon Downtown by a 1755-Like Tsunami." *Nat. Hazards Earth Syst. Sci.*, Vol. 11, pp. 3319–3326.

Bazin, H. (1865). "Recherches Expérimentales sur la Propagation des Ondes." ('Experimental Research on Wave Propagation.') *Mémoires présentés par divers savants à l'Académie des Sciences*, Paris, France, Vol. 19, pp. 495-644 (in French).

Bombardelli, F., and Chanson, H. (2009). "Progress in the Observation and Modeling of Turbulent Multi-Phase Flows." *Environmental Fluid Mechanics*, Vol. 9, No. 2, pp. 121-123 (DOI: 10.1007/s10652-009-9125-8).

Bukreev, V. I. (2004). "On the Critical Velocities and Depths in a Nonuniform Steady Flow in an Open Channel." *Vodnye Resursy*, Vol. 31, No. 1, pp. 40–45. (English translation: *Water Resources*, 2004, Vol. 31, No. 1, pp. 35-40.)

Caputo, J.-G., and Stepanyants, Y. A. (2003). "Bore Formation, Evolution and Disintegration into Solitons in Shallow Inhomogeneous Channels." *Nonlinear Processes in Geophysics*, Vol. 10, pp. 407–424.

Cartwright, J.H.E., and Nakamura, H. (2009). "What Kind of a Wave is Hokusai's

Great Wave off Kanagawa?" *Notes & Records of the Royal Society*, Vol. 69, pp. 119 -135. CHANSON, H. (2004). *"Environmental Hydraulics of Open Channel Flows." Elsevier-Butterworth-Heinemann*, Oxford, UK, 483 pages.

Chanson, H. (2005). "Le Tsunami du 26 Décembre 2004: un Phénomène Hydraulique d'Ampleur Internationale. Premier Constats." ('The 26 December 2004 Tsunami: a Hydraulic Engineering Phenomenon of

International Significance. First Comments') *Journal La Houille Blanche*, No. 2, pp. 25-32 (in French).

Chanson, H. (2009). "Current Knowledge in Hydraulic Jumps and Related Phenomena. A Survey of Experimental Results." *European Journal of Mechanics B/Fluids*, Vol. 28, No. 2, pp. 191-210.

Chanson, H. (2010a). "Coastal Observations: Tsunami Warning Signs on the Enshu Coast of Japan." *Shore & Beach*, Vol. 78, No. 1, pp. 52-54.

Chanson, H. (2010b). "Unsteady Turbulence in Tidal Bores: Effects of Bed Roughness." *Journal of Waterway, Port, Coastal, and Ocean Engineering*, ASCE, Vol. 136, No. 5, pp. 247-256 (DOI: 10.1061/(ASCE)WW.1943-5460.0000048).

Chanson, H. (2010c). "Undular Tidal Bores: Basic Theory and Free-surface Characteristics." *Journal of Hydraulic Engineering*, ASCE, Vol. 136, No. 11, pp. 940-944 (DOI: 10.1061/(ASCE)HY.1943-7900.0000264).

Chanson, H. (2011a). "Tidal Bores, Aegir, Eagre, Mascaret, Pororoca: Theory and Observations." *World Scientific*, Singapore, 220 pages (ISBN 9789814335416).

Chanson, H. (2011b). "Undular Tidal Bores: Effect of Channel Constriction and Bridge Piers." *Environmental Fluid Mechanics*, Vol. 11, No. 4, pp. 385-404.

Chanson, H. (2011c). "Current Knowledge in Tidal bores and their Environmental, Ecological and Cultural Impacts." *Environmental Fluid Mechanics*, Vol. 11, No. 1, pp. 77-98.

Chanson, H. (2012). "Momentum Considerations in Hydraulic Jumps and Bores." *Journal of Irrigation and Drainage Engineering*, ASCE, Vol. 138, No. 4, pp. 382-385 (DOI 10.1061/(ASCE)IR.1943-4774.0000409).

Chanson, H., and Docherty, N. (2012). "Turbulent Velocity Measurements In Open Channel Bores." *European Journal of Mechanics B/Fluids*, Vol. 32, pp. 52–58.

Chanson, H., and Tan, K.K. (2010). "Turbulent Mixing Of Particles Under Tidal Bores: An Experimental Analysis." *Journal of Hydraulic Research*, Vol. 48, No. 5, pp. 641–649.

Chanson, H., and Tan, K.K. (2011). "Dispersion of Fish Eggs under Undular and Breaking Tidal Bores." *Fluid Dynamics & Materials Processing*, Tech Science Press, Vol. 7, No. 4, pp. 403-418.

Chanson, H., Reungoat, D., Simon, B., and Lubin, P. (2011). "High-Frequency Turbulence and Suspended Sediment Concentration Measurements in the Garonne River Tidal Bore." *Estuarine Coastal and Shelf Science*, Vol. 95, No. 2-3, pp. 298-306 (DOI 10.1016/j.ecss.2011.09.012).

Chanson, H., Lubin, P., and Glockner, S. (2012). "Unsteady Turbulence in a Shock: Physical and Numerical Modelling in Tidal Bores and Hydraulic Jumps." in "Turbulence: Theory, Types and Simulation", Nova Science Publishers, Hauppauge NY, USA, Ed. R.J. MARCUSO, Chapter 3, pp. 113-148 (ISBN 978-1-61761-735-5).

Chen, S. (2003). "Tidal Bore In The North Branch Of The Changjiang Estuary." International Conference on Estuaries and Coasts, November 9-11, Hangzhou, China.DELWAR, H.S.M. (2001). "Hydrodynamic Characteristics of the Coastal Areas of Bangladesh."*Proc. Intl Seminar on Quaternary Development and Coastal Hydrodynamics of Ganges Delta Bangladesh* , Khorshed-Alam-A-K-M and Khairul-Islam-M Editors, Geological Survey of Bangladesh Publ., Dhaka, Bangladesh, pp. 69-85.

Didenkulova, I., and Pelinovsky, E. (2011). "Runup of Tsunami Waves in U-Shaped Bays." *Pure Appl. Geophys.*, Vol. 168, pp. 1239–1249.

Docherty, N., and Chanson, H. (2012). "Physical Modeling of Unsteady Turbulence in Breaking Tidal Bores." *Journal of Hydraulic Engineering*, ASCE, Vol. 138, No. 5, pp. 412-419 (DOI: 10.1061/(ASCE)HY.1943-7900.0000542).

Donnelly, C., and Chanson, H. (2005). "Environmental Impact of Undular Tidal Bores in Tropical Rivers." *Environmental Fluid Mechanics*, Vol. 5, No. 5, pp. 481-494 (DOI: 10.1007/s10652-005-0711-0).

Fan, D.D., Cai, G.F., Shang, S., Wu, Y.J., Zhang Y.W., and Gao, L. (2012). "Sedimentation Processes and Sedimentary Characteristics of Tidal Bores along the North Bank of the Qiantang Estuary." *Chinese Science Bulletin*, Vol. 57, No.13, pp. 1578-1589.

Finkl, C. W., Pelinovsky, E. N., and Cathcart, R. B. (2012). "A Review of Potential Tsunami Impacts to the Suez Canal." *Journal of Coastal Research*, Vol. 28, No. 4, pp. 745 - 759.

Fritz, H., Petroff, C., Catala, P., Cienfuegos, R., Winckler, P., Kalligeris, N., Weiss, R., Barrientos, S, Meneses, G., Valderas-Bermejo, C, Ebeling, C., Papadopoulos, A., Contreras, M., Almar, R., Dominguez, J.-C., and Synolakis, C. (2011). "Field Survey of the 27 February 2010 Chile Tsunami." *Pure Appl. Geophys.*, Vol. 168, pp. 1989–2010.

Furgerot, L., Mouaze, D., Tessier, B., and Brun-Cottan, J.C. (2012). "Tidal Bore: Eulerian Velocities and Suspended Sediment Concentration Measurements." *Proc. Intl Conf. on Fluvial Hydraulics River Flow 2012*, San Jose, Costa Rica, 8 pages (CD-ROM).

Furuyama, S., and Chanson, H. (2010). "A Numerical Solution of a Tidal Bore Flow." *Coastal Engineering Journal*, Vol. 52, No. 3, pp. 215-234.

Henderson, F.M. (1966). "Open Channel Flow." *MacMillan Company*, New York, USA. HOKUSAI, K. (1826-1833). "*Thirty-Six Views of Mount Fuji.*" Japan.

Hornung, H.G., Willert, C., and Turner, S. (1995). "The Flow Field downstream of a Hydraulic Jump." *Jl of Fluid Mech.*, Vol. 287, pp. 299–316.

Kayane, K., Tanaka, H., and Adityawan, M.B. (2012). "Influence of River Bed Slope for Tsunami Propagation into Rivers." *Journal of Japan Soc. of Civil Engineers*, Series B2 (Coastal Engineering), Vol. B2-68, 5 pages.

Khezri, N., and Chanson, H. (2012a). "Sediment Inception under Breaking Tidal Bores." *Mechanics Research Communications*, Vol. 41, pp. 49-53 & 1 vide movie (DOI 10.1016/j.mechrescom.2012.02.010).

Khezri, N., and Chanson, H. (2012b). "Inception of Bed Load Motion beneath a Bore." *Geomorphology*, Vol. 153–154, pp. 39–47 & 2 video movies (DOI: 10.1016/j.geomorph.2012.02.006).

Khezri, N., and Chanson, II. (2012c). "Undular and Breaking Bores on Fixed and Movable Gravel Beds." *Journal of Hydraulic Research*, Vol. 50, No. 4, pp. 353–363(DOI: 10.1080/00221686.2012.686200).

Kjerfve, B., and Ferreira, H.O. (1993). "Tidal Bores: First Ever Measurements." *Ciência e Cultura (Journal of the Brazilian Association for the Advancement of Science)*, Vol. 45, No. 2, pp. 135–138.

Koch, C., and Chanson, H. (2008). "Turbulent Mixing beneath an Undular Bore Front." *Journal of Coastal Research*, Vol. 24, No. 4, pp. 999-1007.

Koch, C., and Chanson, H. (2009). "Turbulence Measurements in Positive Surges and Bores." *Journal of Hydraulic Research*, IAHR, Vol. 47, No. 1, pp. 29-40.

Koshimura, S., Matsuoka, M., Matsuyama, M., Yoshii, T., Mas, E., Jimenez, C., and Yamazaki, F. (2011). "Field Survey of the 2010 Tsunami in Chile." *Proceedings of the 8th International Conference on Urban Earthquake Engineering*, Tokyo Institute of Technology, Japan, 7í8 March.

Lacombe, H. (1952). "Quelques Aspects du Problème des Marées Fluviales et de la Formation du Mascaret." ('Some aspects of the Problem of Tides in Rivers and the Formation of a Tidal Bore.') *Bulletin d'information*, Comité Central d'Océanographie et d'Études des Côtes (COEC), France, Vol. 4, No. 6, June, pp. 228-254 (in French).

Liggett, J.A. (1994). "Fluid Mechanics." *McGraw-Hill*, New York, USA.

Lighthill, J. (1978). "Waves in Fluids." *Cambridge University Press*, Cambridge, UK, 504 pages.

Lubin, P., Glockner, S., and Chanson, H. (2010a). "Numerical Simulation of a Weak Breaking Tidal Bore." *Mechanics Research Communications*, Vol. 37, No. 1, pp. 119-121.

Lubin, P., Chanson, H., and Glockner, S. (2010b). "Large Eddy Simulation of Turbulence Generated by a Weak Breaking Tidal Bore." *Environmental Fluid Mechanics*, Vol. 10, No. 5, pp. 587-602.

Madsen, P. A., Simonsen, H. J., and Pan, C.-H. (2005). "Numerical Simulation of Tidal Bores and Hydraulic Jumps." *Coastal Eng.*, Vol. 52, pp. 409-433.

Madsen, P. A., Fuhrman, D. R., and Schäffer, H. A. (2008). "On the Solitary Wave Paradigm for Tsunamis". *J. Geophys. Res.*, Vol. 113, C12012, doi:10.1029/2008JC004932.

Madsen, P. A. (2010). "On the Evolution and Run-up of Tsunamis." *Journal of Hydrodynamics*, Vol. 22, No. 5 (Suppl. 1), pp. 1-6.

Montes, J.S. (1998). "Hydraulics of Open Channel Flow." *ASCE Press*, New-York, USA, 697 pages.

Mori, N., Takahashi, T., Yasuda, T., and Yanagisawa, H. (2011). "Survey of 2011 Tohoku earthquake tsunami inundation and run-up." *Geophysical Research Letters*, Vol. 38, L00G14, 6 pages.

Mouaze, D., Chanson, H., and Simon, B. (2010). "Field Measurements in the Tidal Bore of the Sélune River in the Bay of Mont Saint Michel (September 2010)." *Hydraulic Model Report No. CH81/10*, School of Civil Engineering, The University of Queensland, Brisbane, Australia, 72 pages (ISBN 9781742720210).

Oshiki, H., Chiba, S., Iwase, H., Fujima, K., and Matui, K. (2008). "Numerical Simulation of Tsunami Run-up on the Yoneshiro River by the 1983 Nihonkai-chubu Earthquake." *Annual Jl of Coastal Eng.*, JSCE, Vol. 55, No. 1, pp. 241-245 (in Japanese).

Pan, C., and Huang, W. (2010). "Numerical Modeling of Suspended Sediment Transport Affected by Tidal Bore in Qiantang Estuary." *Jl of Coastal Research*, Vol. 26, No. 6, pp. 1123-1132.

Pelinovsky, E. N., and Rodin, A. A. (2011). "Nonlinear Deformation of a Large Amplitude Wave on Shallow Water." *Doklady Physics*, Vol. 56, No. 5, pp. 305–308.

Poincaré, H. (1910). "Leçons de Mécanique Céleste. Tome 3. Théorie des Marées." ('Lectures on Celest Mechanics. Volume 3. Theory of Tides.') *Gauthier-Villars*, Paris, France (in French).

Rao, G., and Lin, A. (2011). "Distribution of Inundation by the Great Tsunami of the 2011 M_w 9.0 Earthquake off the Pacific Coast of Tohoku (Japan), as

Revealed by ALOS Imagery Data." *International Journal of Remote Sensing*, Vol. 32, No. 22, pp. 7073-7086.

Rayleigh, Lord (1908). "Note on Tidal Bores." *Proc. Royal Soc. of London, Series A containing Papers of a Mathematical and Physical Character*, Vol. 81, No. 541, pp. 448-449.

Reungoat, D., Chanson, H., and Caplain, B. (2012). "Field Measurements in the Tidal Bore of the Garonne River at Arcins (June 2012)." *Hydraulic Model Report No. CH89/12*, School of Civil Engineering, The University of Queensland, Brisbane, Australia, 121 pages (ISBN 9781742720616).

Roh, M., Tanaka, H., Aditaywan, M.B., Mano, A., and Udo, K. (2012)." Evaluation of Celerity and Velocity for Tsunami Propagation into Rivers." *Proc. 33rd Intl. Conf. on Coastal Eng.*, Santander, Spain, P. LYNETT and J. McKEE SMITH Editors, 10 pages.

Rowbotham, F. (1983). "The Severn Bore." *David & Charles*, Newton Abbot, UK, 3rd edition, 104 pages.

Rulifson, R.A., and Tull, K.A. (1999). "Striped Bass Spawning in a Tidal Bore River: the Shubenacadie Estuary, Atlantic Canada." *Transactions of the American Fisheries Society*, Vol. 128, pp. 613-624.

Shuto, N. (1985). "The Nihonkai-Chubu Earthquake Tsunami on the North Akita Coast." *Coastal Eng. in Japan*, Vol. 20, pp. 250-264.

Simpson, J.H., Fisher, N.R., and Wiles, P. (2004). "Reynolds Stress and TKE Production in an Estuary with a Tidal Bore." *Estuarine, Coastal and Shelf Science*, Vol. 60, No. 4, pp. 619-627.

Sommer, A., and Mosley, W.H. (1972). "East Bengal Cyclone of November 1970. Epidemological Approach to Disaster Assessment." *Le Lancet*, Vol. 299, No. 7759, Sat. 13 May, pp. 1029-1036.

Tanaka, H., Ishono, K., Nawarathna, B., Nakagawa, H., Yano, S.-I., Yasuda, H., Watanabe, Y., and Hasegawa, K. (2008). "Field Investigation Of Disasters In Sri Lankan Rivers Caused By Sumatra Earthquake Tsunami." *J. Hydraulics, Hydrology and Environment*, JSCE, 26(1), 91-112.

Tanaka, H., Tinh, N.X., and Dao, N.X. (2011). "Field Measurement and Numerical Studies on the Tsunami Propagation into Upstream Rivers." *Proc. 34th IAHR World Congress*, Brisbane, Australia, 26 June-1 July, Engineers Australia Publication, Eric Valentine, Colin Apelt, James Ball, Hubert Chanson, Ron Cox, Rob Ettema, George Kuczera, Martin Lambert, Bruce Melville and Jane Sargison Editors, pp. 1317-1324.

Tanaka, H., Nguyen, X.T., Umeda, M., Hirao, R., Pradjoko, E., Mano, A., and Udo, K. (2012a). "Coastal and Estuarine Morphology Changes Induced by the 2011 Great East Japan Earthquake Tsunami." *Coastal Engineering*

Journal, Vol. 54, No. 1, paper 1250010, 25 pages (DOI: 10.1142/S0578563412500106).

Tanaka, N., Yagisawa, J., and Yasuda, S. (2012b). "Characteristics of Damage due to Tsunami Propagation in River Channels and Overflow of their Embankments in Great East Japan Earthquake." *International Journal of River Basin Management*, Vol. 10, No. 3, pp. 269-279.

Tinh, N.X., Tanaka, H., Hirao, R., Pradjoko, E., Mano, A., and Udo, K. (2012)." Morphology Change caused by the 2011 Great Tohoku Tsunami Wave - Mechanism of Recent Closure and Breaching of the Nanakita River Mouth." *Proc. 33rd Intl. Conf. on Coastal Eng.*, Santander, Spain, P. LYNETT and J. Mckee SMITH Editors, 10 pages.

Tsuji, Y., Yanuma, T., Murata, I., and Fujiwara, C. (1991). "Tsunami Ascending in Rivers as an Undular Bore." *Natural Hazards*, Vol. 4, pp. 257-266.

Tessier, B., and Terwindt, J.H.J. (1994). "An Example of Soft-Sediment Deformations in an intertidal Environment - The Effect of a Tidal Bore". *Comptes-Rendus de l'Académie des Sciences*, Série II, Vol. 319, No. 2, Part 2, pp. 217-233 (in French).

Wijetunge, J. J., Wang, X., and Liu, P. L.-F., (2008). "Indian Ocean Tsunami on 26 December 2004: Numerical Modeling of Inundation in Three Cities on the South Coast of Sri Lanka." *Journal of Earthquake and Tsunami*, Vol. 2, No. 2, pp. 133-155.

Wijetunge, J. J. (2009a). "Field Measurements and Numerical Simulations of the 2004 Tsunami Impact on the East Coast of Sri Lanka." *Pure Appl. Geophys*., Vol. 166, pp. 593–622.

Wijetunge, J. J. (2009b). "Field Measurements and Numerical Simulations of the 2004 Tsunami Impact on the South Coast of Sri Lanka." *Ocean Engineering*, Vol. 36, pp. 960–973.

Wolanski, E., Moore, K., Spagnol, S., D'adamo, N., and Pattieratchi, C. (2001). "Rapid, Human-Induced Siltation of the Macro -Tidal Ord River Estuary, Western Australia." *Estuarine, Coastal and Shelf Science*, Vol. 53, pp. 717-732.

Wolanski, E., Williams, D., Spagnol, S., and Chanson, H. (2004). "Undular Tidal Bore Dynamics in the Daly Estuary, Northern Australia." *Estuarine, Coastal and Shelf Science*, Vol. 60, No. 4, pp. 629-636.

Wolanski, E., Williams, D., and Hanert, E. (2006). "The Sediment Trapping Efficiency of the macro-tidal Daly Estuary, Tropical Australia." *Estuarine, Coastal and Shelf Science*, Vol. 69, pp. 291-298.

Yasuda, H., Watanabe, Y., and Fujima, K. (2004). "Report on River Propagation of Tsunami Generated by the Tokachi-Oki Earthquake on 26 September, 2003." *J. Hydraulics, Coastal and Environment Engineering*, JSCE, No. 768/11-68, 209-218 (in Japanese with English abstract).

Yasuda, H. (2010). "One-Dimensional Study on Propagation of Tsunami Wave in River Channels." *Jl of Hydraulic Engineering*, ASCE, Vol. 136, No. 2, pp. 93-105.

Yeh, H., Tolkova, E., Jay, D., Talke, S., and Fritz, H. (2012). "Tsunami Hydrodynamics in the Columbia River." *Journal of Disaster Research*, Vol.7, No. 5, pp. 604-608.

In: Tsunamis
Editor: Tianxing Cai
ISBN: 978-1-62808-682-9

Chapter 4

"What - Me Worry[1]?" "Why so Serious[2]?" A Personal View on the Fukushima Nuclear Reactor Accidents[3]

Raymond H. V. Gallucci
US Nuclear Regulatory Commission

Abstract

Infrequently, it seems that a significant accident precursor or, worse, an actual accident, involving a commercial nuclear power reactor occurs to remind us of the need to re-examine the safety of this important electrical power technology from a risk perspective. Twenty-five years since the major core damage accident at Chernobyl in the Ukraine, the Fukushima reactor complex in Japan experienced multiple core damages as a result of an earthquake-induced tsunami beyond either the earthquake or tsunami design basis for the site. While the tsunami itself killed tens of thousands of people and left the area devastated and virtually

---

[1] Alfred E. Neuman, Mad Magazine

[2] Heath Ledger as "The Joker," in The Dark Knight.

[3] This paper was written shortly after the earthquake and tsunami that devastated the Fukushima nuclear complex in Japan and surrounding region as the author's attempt to recover "perspective" in the wake of the resulting nuclear plant accidents. It was originally published under the same title in Risk Analysis: An International Journal, Volume 32, Number 9, Society for Risk Analysis, McLean, Virginia, September 2012, pp. 1444-1450.

uninhabitable, much concern still arose from the potential radioactive releases from the damaged reactors, even though there was little population left in the area to be affected. As a lifelong probabilistic safety analyst in nuclear engineering, even I must admit to a recurrence of the doubt regarding nuclear power safety following Fukushima that I had experienced following Three Mile Island and Chernobyl. This article is my attempt to "recover" my personal perspective on acceptable risk by examining both the domestic and worldwide history of commercial nuclear power plant accidents and attempting to quantify the risk in terms of the frequency of core damage that one might glean from a review of operational history.

INTRODUCTION

Since the commercial nuclear industry last received a major wake-up call at Chernobyl in 1986, there have been some "near-misses" (e.g., Davis-Besse in 2002 [near loss-of-coolant accident due to corrosion of the reactor pressure vessel head]). However, nothing had even approached the level of Chernobyl until the earthquake-tsunami disaster at Fukushima in 2011, the first ever multi-unit set of core damage accidents. It was also the first ever core damage(s) caused by an "external event" (multi-system to plant-wide common-cause failures due to phenomena such as fires or earthquakes), despite the plants' design bases against earthquakes and tsunamis. Nuclear power opponents have been rejuvenated, not unlike when extrapolating the terrorism of September 11, 2001, to nuclear plants (e.g., see Appendix). And, despite being a lifelong nuclear engineer with over thirty years in probabilistic safety analysis (PSA), even I must admit to a recurrence of the doubt similar to what I felt following Three Mile Island (1979) and Chernobyl. At least with these two accidents, I could take "comfort" in knowing that both involved significant human errors that should not have occurred and have since been corrected. I base- this on the nuclear industry's safety record with respect to "internal events" (equipment failures or human errors during "normal" operation) since Chernobyl (Davis-Besse in 2002 notwithstanding). Can I return to my previous "comfort level" with respect to external events in light of not so much the earthquake (beyond the design basis, but apparently still unable to compromise fuel integrity by itself), but the tsunami (which ultimately caused the three core damages, in addition to radioactive releases from the spent fuel pools)?

QUESTIONS

As a PSA practitioner, I often take solace in my trade, given the ability to model both equipment and human interactions in great detail for nuclear plants, at least for "internal events." Complex system interactions, including humans, are modeled to the Nth degree, including common-cause failures, providing reassurance, perhaps somewhat naïvely, that we thoroughly understand our reactors. At least in the US, many regulatory applications now take advantage of this modeling ability such that former reliance on what have been deemed overly conservative, deterministic mitigation measures for "worst case" accidents have been supplemented, if not replaced, by reliance on the perceived rarity (virtual impossibility?) of such worst-case accidents. However, might I be somewhat naïve when believing we can extrapolate the detailed modeling we use for internal-event-related core damage to that due to external events? Is this especially true when considering earthquakes or tsunamis that cause widespread devastation and kill thousands exclusive of any potential compounding from reactor accidents? Is the implicit uncertainty involved in the complex phenomenology of such events as fire, flood, extreme wind/storm, earthquake, or tsunami beyond our ability to believably model? Should I be worried? Should I take such threats seriously enough to try to defend against them, assuming I even can? I explore this, not at the more common detailed level of PSA, but at the higher level of statistical analysis of the operating history of commercial nuclear power.

QUANTIFYING RADIOLOGICAL EVENTS

The International Atomic Energy Agency (IAEA) classifies nuclear and radiological events at seven levels of the International Nuclear and Radiological Event Scale (INES): Levels 4-7 are termed "accidents" and Levels 1-3 "incidents." [1] Events without safety significance are classified as "Below Scale/Level 0." For commercial nuclear plants, accidents leading to core damage start at Level 4, defined as involving *"Fuel melt or damage to fuel* resulting in more than 0.1% release of core inventory." Level 3 corresponds to "close calls" or "near misses" that compromise defense-in-depth but do not cause fuel damage. Therefore, Level 4 seems a reasonable threshold for identifying "core damage accidents" that have occurred worldwide. Since I am considering "only" the occurrence of core damage and not the degree of damage, all accidents at Level 4 and above are viewed as equivalent. For example, the worst Level 7 accidents, at Chernobyl and Fukushima, are viewed the same in terms of the occurrence of core damage as the less severe accidents, such as the Level 5 at Three Mile Island.

I acknowledge that focusing "only" on core damage limits my analysis, and my perspective, to what one might consider a simplistic level. Clearly there are other consequences of concern beyond "just" core damage, such as acute and chronic health effects, land contamination and rehabitability, and economic impacts. I intentionally limit my discussion and analysis to that of core damage, for which the most widespread use of quantitative measures has been applied worldwide and in regulatory applications. Perspectives based on other consequences could be quite different from that which I offer.

ESTIMATING CORE DAMAGE FREQUENCY

Historically, there have been four accidents at commercial nuclear plants of Level 4 or higher:

1) Three Mile Island in 1979 in Middletown, Pennsylvania (US), where equipment failures and operator errors caused a loss of coolant and partial core meltdown, with offsite radiological release limited to 10 μSv (1 mrem);
2) St. Laurent des Eaux in 1980 in Orleans, France, where one channel of fuel in the reactor ruptured due to a brief power excursion, with no offsite radioactive release;
3) Chernobyl in 1986 in Prypiat, Ukraine (then USSR), where an uncontrolled power excursion caused a steam explosion and extensive core meltdown, radioactively contaminating the surrounding area and causing ~50 documented fatalities, mostly to cleanup personnel; and
4) Fukushima (three reactors) in 2011 near Sendai, Japan, where an offshore earthquake of Richter magnitude 9.0 triggered a beyond-design-basis tsunami that caused complete and extended station blackout with ensuing damage to the reactor cores and spent fuel pools due to loss of cooling. [2,3]

As Level 7 accidents, Fukushima is more severe than Three Mile Island, including possible damage to and release from up to four spent fuel pools. [4] Therefore, while recognizing the spectrum of core damage over these four accidents is rather wide, it seems reasonable to assume that, worldwide, there have now been six core damage accidents, counting three from Fukushima. Statistically, given ~14,000 reactor-years (ry) of commercial power plant operation worldwide, this translates into a simple point estimate for core damage frequency (CDF) of ~4E-4/ry. [5] This does not include possible contributions from "fractional core damages" due to "near misses" such as the Browns Ferry Fire in 1975 in Decatur, Alabama (US), for which an estimate of 0.2 core damage has been made by the US Nuclear Regulatory Commission (USNRC) [6]. I now consider some refinements to my simple statistical estimate for CDF.

First Refinement

First, I observe that, of the six accidents, three could be considered due solely to "internal events" (Three Mile Island, Saint Laurent des Eaux and Chernobyl), and three to "external events" (Fukushima). For the internal core damage events, none has occurred since Chernobyl in 1986, suggesting that improvements made worldwide since, say, five years after Chernobyl (i.e., starting in 1992), have reduced the CDF from internal events[4]. From 1992 through now, there have been ~8,000 ry of operation worldwide, suggesting a CDF from a "zero-failure-rate (ZFR) approximation," applicable since no internal event core damage has occurred since 1992, based on the 50th percentile upper bound of a Poisson distribution, of 0.7/(8000 ry) = 9E-5/ry. [5,7] Next, consider that there is no evidence for a similar reduction in the CDF due to external events, at least for earthquakes and consequential tsunamis, over the world operating history. Even excluding the potential contribution from "fractional core damages" due to "near misses" such as Browns Ferry[5], this translates into a simple point estimate for CDF due to external events of ~3/(14000 ry) = 2E-4/ry. [8] Combined with the preceding, we postulate at least one rather simplistic point estimate for total CDF (internal and external events) of 9E-5/ry + 2E-4/ry ≈ 3E-4/ry worldwide.

[4] Some of these improvements began even earlier, especially in the US, as a result of lessons learned from Browns Ferry, an "external event" for which enhancements provided benefits "internally" as well, and Three Mile Island.

[5] And one might argue this exclusion is justified given improvements made since Browns Ferry and the fact that the last "significant" fire "incident," not "accident," at a commercial plant occurred in 1993 at Narora in India, where the control room had to be evacuated.

Second Refinement

Now, I attempt to refine these estimates using the more comprehensive evaluation of conditional core damage probabilities (CCDPs) performed by the USNRC for commercial nuclear plant events in the US via the Accident Sequence Precursor (ASP) Program[6]. The ASP "systematically evaluates US nuclear power plant operating experience to identify, document, and rank the operating events most likely to lead to inadequate core cooling and severe core damage (precursors)." [6] A list of all "significant" precursors that have occurred since 1969, i.e., events where the estimated "fractional core damage," i.e., CCDP, was ≥0.001, indicates a cumulative CCDP of ~1.5, of which ~90% arises from the five highest CCDPs ranging from 1.0 for Three Mile Island to 0.02 during an at-power test that failed all three auxiliary feedwater pumps at Turkey Point in 1974. Of the 34 events evaluated, only one, Browns Ferry in 1975, can be attributed to an external event (fire), with the estimated CCDP ≈ 0.2. Recognizing that US commercial plants have accumulated ~3,400 ry of operation, a simple point estimate for the total CDF becomes ~1.5/(3400 ry) = 4.4E-4/ry, about the same as the initial estimate we made for plants worldwide (i.e., 6/[14000 ry] ≈ 4E-4/ry). [9] However, whereas worldwide this simple estimate was split equally between internal and external events (three internal event and three external event core damages), for the US the contribution arises mostly from internal events, as a result of the dominance from Three Mile Island (~1.3/[3400 ry] = 3.8E-4/ry from internal and ~0.2/[3400 ry] = 6E-5/ry from external).

This estimate is based on the entire US commercial operating history and implicitly includes conditions both before and after Browns Ferry and Three Mile Island, i.e., it is a combination of what might be expected to include both "unenhanced" and "enhanced" safety conditions at US plants. If I assume significant safety enhancements were made subsequent to Browns Ferry and Three Mile Island, say, roughly five years after occurrence of the latter (or starting in 1985), this reduces the cumulative CCDP to only 0.033, arising from eight events, none later than the Davis-Besse near-miss in 2002. These events had CCDPs ranging from 0.01 at Davis-Besse in 1985 (the 2002 event was assigned a lower value of 0.006) to 0.001 at Turkey Point in 1986. The

[6] CCDP is the probability that, given the occurrence of an accident initiator (i.e., the "condition"), core damage will result from failures of systems, structures, components and human actions intended to prevent the core damage. In effect, it is a measure of the "fraction of core damage," recognizing that the actual occurrence of core damage, such as occurred at Three Mile Island, is unity.

1985 Davis-Besse event resulted from an at-power trip of one main feedwater pump while the second was unavailable, requiring subsequent closure of the block valve for the stuck-open power-operated relief valve. The 1986 Turkey Point event resulted from a manual trip following loss of the turbine governor oil system and a resulting partial failure to scram. None of these eight events can be associated with an external event. Given ~2,900 ry of operation in the US from 1985 onward, a much reduced estimate for total CDF, arising exclusively from internal events, becomes ~0.033/(2900 ry) = 1.1E-5/ry, or a factor of ~40 lower than the previous estimate, which included both the contributions from Three Mile Island and Browns Ferry[7].

Worldwide vs. US

So far, my very simplified examination seems to suggest a dichotomy between internal and external event CDFs worldwide vs. domestically, as follows. If I consider the entire operating histories only, one first sees essentially equal "total" CDFs of ~4E-4/ry (~6/[14000 ry] worldwide and ~1.5/[3400 ry] in the US)[8]. However, parsing these into internal and external CDFs, one sees essentially equal values worldwide, each ~2E-4/ry (i.e., 3/[14000 ry]), while the corresponding values for the US are 3.8E-4/ry and 6E-5/ry, respectively. Additional reductions result if one assumes that there have been significant enhancements, both worldwide and domestically, to reduce internal CDF as a result of lessons learned from the three "internal" core damages that have occurred (Three Mile Island in 1979, St. Laurent des Eaux in 1980 and Chernobyl in 1986), such that no internal-event-induced core damage has occurred since Chernobyl. For a worldwide operating period from 1992 onward and a domestic one from 1985 onward, estimates for internal CDF reduce to ~9E-5/ry (via a ZFR approximation) and 1.1E-5/ry (based on

[7] Just coincidentally, if I apply the ZFR approximation approach here for external events, I estimate a CDF about the same as that for external events worldwide, i.e., ~0.7/(2900 ry) = 2E-4/ry. However, on the surface it seems inconsistent to attribute a higher fractional core damage when there has been no near-miss (0.7 external-event-induced core damage since 1985) than when there is a measurable fraction of near-misses (0.033 internal-event-induced core damage since 1985). Should the "coincidence" be dismissed?

[8] The results are necessarily limited by the fact that the CCDP-type estimates used in the "Second Refinement" are consistently available only for US plants via the ASP Program. The availability of similar estimates for non-US plants on a consistent basis would enable the worldwide estimates to be made directly, rather than by extrapolation. Implicit in this extrapolation is a simplifying assumption that safety measures for nuclear power plants outside the US are, on average, comparable to those in the US.

CCDPs from the ASP Program), respectively. While the former is "only" about half its previous corresponding value, the latter shows a reduction by a factor of ~40 relative to its previous corresponding value.

When external events are considered, there clearly is no basis for parsing the operating history below the full 14,000 ry worldwide, given the recent Fukushima accident, so the previous estimate of ~2E-4/ry seems reasonable. In the US, there has been no externally-induced near-miss since Browns Ferry in 1975, and it is known that all plants significantly reduced their vulnerability to fire starting in the 1980s. However, does the fact that the US has not experienced a seismic-related (or other non-fire-external-event-related) near-miss justify assigning a different external event CDF from that worldwide?

"Scaled" Results

For the US, the ASP Program results indicate ~1.5 core damages have occurred, arising from one actual core damage (Three Mile Island) and a variety of "near-misses," such as Browns Ferry. This could suggest a potential way to "scale" historic results for known core damage accidents by increasing the "total" number of core damages by about half based on the number known to have occurred. Worldwide, this would suggest that the five known core damage accidents at non-US commercial nuclear plants could be scaled to 7.5 to account for "near-misses," reported and possibly unreported. While INES levels have not been assigned to all, there have been several other potential "near-misses" worldwide which, if quantified in a manner similar to that for US "near-misses," might account for the estimated additional 2.5 non-US core damages worldwide: (1) melting, and subsequent burning, of a fuel element at the Chapelcross reactor in Scotland in 1967 due to debris blockage of a fuel channel; (2) damage to ten fuel rods at the Greifswald reactor in (East) Germany in 1975 when three of six cooling pumps were shut down for testing and a fourth subsequently failed, causing excessive heating in the core; (3) damage to fuel integrity, along with extensive cladding corrosion and onsite radioactive release, at the Bohunice station in Czechoslovakia in 1977 due to operator failure to remove foreign material (moisture absorber) prior to loading a fuel assembly; (4) a large cable fire in the turbine building at the Beloyarsk reactor in the USSR in 1978 that severely damaged the control building, main control room panels, and redundant safety trains; (5) a large cable gallery fire at the Armenia station in the USSR in 1982 that severely impacted core cooling capability, resulting in station blackout and severed power sources to equipment throughout the plant; (6) loss of safety systems due to fire in a turbo-generator at the Vandellos plant in Spain in 1989 that damaged a pipe expansion joint, leading to flooding of the turbine and auxiliary buildings; (7) a large turbine building fire due to a turbine blade failure at the Narora station in India in 1993, causing station blackout and forcing control room abandonment; (8) a 15-minute reactivity excursion at the Ishikawa station in Japan in 1999 due to operators erroneously removing instead of inserting a control rod during an inspection. [2, 3, 8, 10]

Combined the scaled estimate worldwide of 7.5 core damages with ~1.5 US "core damages" yields nine, or a worldwide total CDF of ~9/(14000 ry) = 6E-4/ry, not a substantial increase over (i.e., "only" 50% higher than) that previously estimated (~4E-4/ry) for the full 14,000-ry operating history based on the six core damage accidents that actually occurred. If I consider that there has been no internal-event-related core damage accident since Chernobyl in 1986, one can again consider internal and external CDFs separately. However, since the worldwide internal CDF was previously based on a ZFR approximation, there would be no change here, since it is conceivable that the 0.7 core damage estimated from that approximation may be a reasonable surrogate for the internal "near-misses" since 1992 (five years after Chernobyl). Thus, one retains the previous internal CDF estimate of 0.7/(8000 ry) ≈ 9E-5/ry. I then scale the worldwide external by 50%, implying an increase to (3 x 1.5)/(14000 ry) ≈ 3E-4/ry. Again, this is not a substantial increase over the previous estimate.

Combined Results

Table 1 summarizes the results obtained so far. For internal-event-related core damage accidents, the CDF estimates range from a high of 3.8E-4/ry, based on the ASP results for the US over the full 3,400-ry operating history, to a low of 1.1E-5/ry, based on the ASP results for the US from 1985 onward (i.e., 2,900 ry, starting five years after Three Mile Island). The worldwide high and low internal CDF estimates are 3E-4/ry and 9E-5/ry, respectively, the first based on scaling via the US ASP results over the full 14,000-ry operating history and the second based on the ZFR approximation for the 8,000 ry that have accumulated since five years after Chernobyl (i.e., starting in 1992). Is it reasonable to expect roughly an order-of-magnitude difference between internal CDF worldwide and domestically (i.e., 3E-4/ry or 9E-5/ry [worldwide] vs. 1.1E-5/ry [US])? Probably not, since there has been no internal-event-related core damage since Chernobyl. Based on the estimates here, I suggest a reasonable range for internal CDF to be between the lows both worldwide and domestic, i.e., ~9E-5/ry and ~1E-5/ry, with the domestic point estimate being lower, say, the geometric mean (~3E-5/ry), and the worldwide point estimate being higher, say, the arithmetic mean (~5E-5/ry).

What about external-event-related core damage accidents? The CDF estimates range from a high of ~3E-4/ry, based on the ASP-scaled worldwide results for the full 14,000-ry operating history, to a low of ~6E-5/ry, based on the Browns Ferry fire and full US operating history of 3,400 ry. While the US has experienced only one externally-induced "near-miss," the Browns Ferry fire, over 35 years ago, the recent Fukushima accidents suggest that at least the seismic (and tsunami) vulnerability remains, to which the US is not immune. While the US west coast is more seismically active and coastal reactors there are more susceptible to tsunamis (but designed to resist these, at least at some design basis), one should not (cannot?) *a priori* dismiss the potential for tsunamis of sufficient magnitude along the east coast that could produce Fukushima-like damage. For example, while experts debate the possible magnitude of a tsunami along the US east coast from collapse of the Cumbre Vieja slope on the volcanic island of La Palma in the Canary Islands[9], certainly a tsunami of magnitude similar to that which struck Fukushima (approaching 50 ft) cannot be dismissed. [11]

Inland reactors, while not susceptible to oceanic tsunamis, remain susceptible to earthquakes that may be higher than their design basis (Fukushima was designed for a 7.5 Richter magnitude earthquake, but experienced one of magnitude 9.0, for which it still was able to shut down and initiate emergency cooling, at least until the tsunami hit).

Therefore, taking the range from ~3E-4/ry to ~6E-5/ry as reasonable for external-event-related CDFs, I suggest point estimates the same both worldwide and domestically, with the inland point estimate (tsunami-free) being lower, say, the geometric mean (~1E-4/ry), and the coastal point estimate (tsunami-susceptible) being higher, say, the arithmetic mean (~2E-4/ry). These suggested values are summarized in Table 2.

[9] There lies on La Palma, Canary, / volcano that renders us wary. / For should it implode / and land slide below / the ocean, our future's quite scary.

Would generate mega-tsunami / from Africa to the Bahamas. / Perhaps hundred feet / of ocean would greet / our coastline from Maine to Miami.

The mountain's named Cumbre Vieja, / volcano that's waiting to slay ya'. / So it should behoove / you pack up and move / to somewhere in west Pennsylvania. (Gallucci R, 2005©).

Table 1. Core Damage Estimates from Internal and External Events Worldwide and in the US

Estimate Type		# of Core Damages {via ASP scaling}[i]	Operating Time (ry)	CDF (1/ry) {via ASP scaling}	Comments
World	Internal	3 {4.5}	14,000	2E-4 {3E-4}	
		0.7	8,000	9E-5	Based on ZFR approximation
	External	3 {4.5}	14,000	2E-4 {3E-4}	
	Total	6 {9}	14,000	4E-4 {6E-4}	Via 14,000 ry for both
		N/A	N/A	3E-4	External + ZFR for internal
US	Internal	1.3	3,400	3.8E-4	
		0.033	2,900	1.1E-5	Since 1985
	External	0.2	3,400	6E-5	Browns Ferry
		0.7	2,900	2E-4	Based on ZFR (since 1985)
	Total	1.5	3,400	4.4E-4	Via 14,000 ry for both
		0.733	2,900	2.5E-4	Internal + ZFR for external (both since 1985), but assigning higher value to non-existent external "near-misses" than actual internal ones

[i] Values in { } result by multiplying original values by 1.5, as discussed in "'Scaled' Results.".

Table 2. Suggested Ranges and Point Estimates for Core Damage Frequencies

Estimate Type			CDF (1/ry) Range	CDF (1/ry) Point Estimate
Internal	World		9E-5 to 1E-5	5E-5
	US			3E-5
External	Inland		3E-4 to 6E-5	1E-4
	Coastal			2E-4
Total	World	Inland	4E-4 to 7E-5	2E-4
		Coastal		3E-4
	US	Inland		1E-4
		Coastal		2E-4

CONCLUSION

Not surprisingly, the results show that external events dominate the total CDF, especially for coastal plants[10]. What does all this tell me? We seem to have a reasonable handle on the potential vulnerabilities due to internal events, as evidenced by the CDF due to these being the minority contributor to total CDF, both worldwide and domestically. Even if we, rightly or wrongly, consider our US plants to be the safest in the world, worldwide operating history seems to support that internal-event-related core damage is less frequent than that due to external events. However, what about these external events? I traditionally have taken comfort that there has never been core damage due to fire worldwide at commercial reactors ("near-misses," yes, but no core damage). [8, 10] And, up to 2011, I could feel a similar comfort with regard to the other external events, such as flood, hurricane, earthquake or tsunami. However, when a country as seismically and "tsunamically" sophisticated as Japan can experience core damages due to one of these events, is any other country, including the supposedly "safer" US, equally "at risk?"

To me, that is where the truth lies – "at risk." The earthquake-induced tsunami at Fukushima was beyond the design basis, and it likely killed over 20,000 and devastated the entire region without any contribution from the

[10] Note that some plants may be located along coasts but be at high enough elevations to be considered "inland" with respect to tsunami vulnerability.

reactor accidents. Certainly it is not as unlikely as the classic doomsday scenario of a meteor strike[11] (which can induce tsunamis beyond any imaginable from an oceanic earthquake or collapse of Cumbre Vieja). Nonetheless, is not the devastation so complete to the local region such that any compounding from radioactive release, even a full "China Syndrome" core melt (which did not occur), would be irrelevant? As bad as the Fukushima core damages have been, they pale in comparison to what the tsunami itself did. My bottom line is that, unless I choose to try to defend against catastrophes on the magnitude of such tsunamis, I need not try to make reactors invulnerable to such events. A Richter magnitude 10 earthquake may be theoretically possible anywhere on earth and, if occurring at sea, capable of generating tsunamis hundreds of feet high. I cannot protect anything against these, not even nuclear plants. My only defense would be never to live near enough to a coast at an elevation below the height of such a tsunami. Since it cannot be resisted, the only solution is total avoidance.

Therefore, despite what happened at Fukushima, I choose not to worry or take such events seriously enough to try to establish what, to me, would be a futile invulnerability against them. We should continue to live along coasts, build nuclear power plants near major cooling sources like the ocean, and accept the risk that comes with these choices.

REFERENCES

[1] International Atomic Energy Agency (IAEA). *The International Nuclear and Radiological Event Scale (INES) User's Manual (2008 Edition)*. Vienna, Austria: 2009.

[2] Guardian News and Media. *Nuclear Power Plant Accidents: Listed and Ranked Since 1952*. Available at: http://www.guardian.co.uk/news/datablog/2011/mar/14/nuclear-power-plant-accidents-list-rank, Accessed on March 22, 2011.

[3] Wikipedia, the Free Encyclopedia. *List of Civilian Nuclear Accidents*. Available at: http://en.wikipedia.org/wiki/ List_of_civilian_nuclear_accidents, Accessed on March 22, 2011.

[11] There may be a growing school of thought questioning the presumed extreme rarity of these. [12].

[4] Wikipedia, the Free Encyclopedia. *Fukushima I Nuclear Accidents*. Available at: http://en.wikipedia.org/wiki/Fukushima_I_nuclear_accidents, Accessed on March 22, 2011.

[5] World Nuclear Association. *Safety of Nuclear Power Reactors*. Available at: http://www.world-nuclear.org/info/info06.html, Accessed on March 22, 2011.

[6] US Nuclear Regulatory Commission (NRC). *Status of the Accident Sequence Precursor Program and the Development of Standardized Plant Analysis Risk Models*. Washington, DC: SECY-10-0125, 2010.

[7] Green A, Bourne A. *Reliability Technology*. New York, NY: Wiley-Interscience, 1972.

[8] Nowlen S, Kazarians M, Wyant F. Risk Methods Insights Gained from Fire Incidents. Washington, DC: NUREG/CR-6738, 2001.

[9] USNRC. *2009-2010 Information Digest*. Washington, DC: NUREG-1350, Volume 21, 2010.

[10] Gallucci R. *Predicting Fire-Induced Core Damage Frequencies – A Simple "Sanity Check."* Transactions of the American Nuclear Society, 2006: Vol. 94.

[11] Ward S, Day S. Cumbre Vieja volcano - *Potential Collapse and Tsunami at La Palma, Canary Islands*. Geophysical Research Letters, 2001: Issue 28-17. Available at: http://www.es.ucsc.edu/~ward/papers/La_Palma_grl.pdf, Accessed on March 22, 2011.

[12] Blakeslee S. *Ancient Crash, Epic Wave*. New York Times, November 14, 2006. Available at: http://www.nytimes.com/ 2006/11/14/science/14WAVE.html, Accessed on March 22, 2011.

[13] World Nuclear News. *US Support for Nuclear Power Remains High*. Available at:http://www.world-nuclear-news.org/NP-US_support_for_nuclear_power_ remains_high-0410114.html, Accessed on October 12, 2011.

[14] Miller C, et al. *Recommendations for Enhancing Reactor Safety in the 21st Century – The Near-Term Task Force Review of Insights from the Fukushima Dai-Ichi Accident*. USNRC, Washington, DC: 2011.

[15] USNRC. *Staff Requirements – SECY-11-0093 – Near-Term Report and Recommendations for Agency Actions Following the Events in Japan*. Washington, DC: 2011.

[16] Reaney P. *Global Support for Nuclear Energy Drops After Fukushima*. Available at: http://www.reuters.com/article/ 2011/06/22/us-energy-nuclear-poll-idUSTRE75L4BY20110622, Accessed on October 12, 2011.

[17] Dalton D. *Fukushima Has Had "No Impact" on UK Public Opinion, Says Poll.* Available at:http://www.worldnuclear.org/ _news_database /rss_detail_ features.cfm? objID=D1DC4A86-68E9-4B7D-991128BD07 9800E5, Accessed on October 12, 2011.
[18] IAEA. *IAEA International Fact Finding Expert Mission of the Fukushima Dai-Ichi NPP Accident Following the Great East Japan Earthquake and Tsunami.* Vienna, Austria: 2011.

APPENDIX

Since I first developed my perspective shortly after the Fukushima accident, there has been time for the public, industry and regulatory response to stabilize somewhat. In the US, a telephone survey of 1000 citizens on behalf of the Nuclear Energy Institute in September 2011 indicates US support for nuclear power remains high. The poll found that 62% still favored nuclear energy as one way to generate electricity, a decrease of only 9% since a similar poll in February 2011, one month before Fukushima. [13] Two-thirds of those polled also rated US nuclear power plant safety as "high," the same as in the February poll.

In July 2011, the USNRC issued the report by its Near-Term Task Force to Review Insights from Fukushima. [14] The Task Force concluded as follows:

> [A] sequence of events like the Fukushima accident is unlikely to occur in the United States and some appropriate mitigation measures have been implemented, reducing the likelihood of core damage and radiological releases. Therefore, continued operation and continued licensing activities do not pose an imminent risk to public health and safety. However, ... a more balanced application of the Commission's defense-in-depth philosophy using risk insights would provide an enhanced regulatory framework that is logical, systematic, coherent, and better understood.

Twelve specific recommendations were offered in the following five categories:

1) Clarifying the Regulatory Framework;
2) Ensuring Protection;
3) Enhancing Mitigation;

4) Strengthening Emergency Preparedness; and
5) Improving the Efficiency of USNRC Programs.

The Commission subsequently instructed USNRC staff to proceed with stakeholder interactions to review and assess these recommendations to develop a "charter for the structure, scope and expectations for assessing the Task Force recommendations and NRC's longer term review." [15]

Not surprisingly, the short-term reaction of the international community to Fukushima was more pessimistic than that of the US. A poll of nearly 19,000 people in 24 countries in June 2011 indicated support had dropped significantly, with 62% worldwide now opposed to nuclear power. [16] Nonetheless, at least in the United Kingdom (UK), the longer term effect of Fukushima has been minimal. A poll there in September 2011 indicted a 3% increase from 2010 to 41% in respondents agreeing that the benefits of nuclear power outweigh the risks. [17] Another increase of 14% from 2005 to 23% has occurred in respondents favoring the UK building more nuclear power plants. Still, neither of these constitutes a majority.

The international community responded promptly to the Fukushima accident through the IAEA, which issued the report by its International Fact Finding Expert Mission in June 2011. [18] Fifteen conclusions and 16 lessons resulted from the Mission, among which are the following:

1) Defense-in-depth against tsunamis was insufficient;
2) Regulatory requirements and guidelines should be updated, including consideration of the compound effects for sites with multiple units;
3) For severe accidents, simple alternative mitigation sources should be available (e.g., mobile electric power, air and water supplies);
4) The potential for external events, such as earthquakes or tsunamis, to affect multiple units at one site requires that a large resource of trained responders, equipment, supplies, etc., be available on short notice;
5) Hydrogen explosion risks and implications should be reassessed; and
6) "Deliberate evacuation" and "evacuation-prepared areas" are more effective in reducing radiation exposure than long-term sheltering.

Clearly there has been, and likely will continue to be, a comprehensive response to the Fukushima accident, beyond what I would consider necessary in light of my assessment. Whether or not this ultimately proves to be more form or substance remains to be seen.

In: Tsunamis
Editor: Tianxing Cai
ISBN: 978-1-62808-682-9

Chapter 5

FUTURE ROLE OF MATHEMATICAL MODELING FOR ECONOMIC ANALYSIS AND ENERGY DISPATCH OPTIMIZATION UNDER EMERGENCY OF TSUNAMIS

Tianxing Cai*
Dan. F Smith Department of Chemical Engineering, Lamar University, Texas, US

ABSTRACT

The natural catastrophe such as tsunamis, earth quake and hurricane will always bring a blow to local economy. The short-time energy shortage is one of the most serious damages in the disaster because it may cause local areas to suffer from delayed rescues, widespread power outages, tremendous economic losses, and even public safety threats.

In such situation, agile energy dispatching through an effective energy transportation network, targeting the minimum energy recovery time, should be a top priority.Therefore, the application of computer aid design will be very essential to handle the above global concern. The mathematical modeling of economic optimization and dispatch scheduling will provide sufficient technical support to decision-makers. The computer simulation and optimization should provide optimal output

* 4400 MLK Blvd, Dan.F Smith Department of Chemical Engineering, Lamar University, Email:tcai@lamar.edu.

solution with the huge input information of variable demand, local energy resources, energy facility capacity, economic sustainability and emergency under extreme conditions. Available renewable technology has provided more potential sources for local energy recovery while it has added the difficulty in the mathematical modeling of design and dispatch of multiple energy types in a wide regional area. In this chapter, the latest progress in this research area will be introduced and reviewed.

Furthermore, our newly developed mathematical models will also be introduced. It is developed for energy network optimization for the consideration of both energy recovery capability and economic factors. At first, potential energy network needs to be characterized, where the capacity, quantity, and availability of various energy sources are put into the quantification. Then, the energy initial situation under emergency conditions needs to be identified. Next step, the energy dispatch requirement is represented as constraint based on a developed MILP (mixed-integer linear programming) model. Followed by the above step, mathematical modeling with the objective function of minimal operation cost is constructed. Finally, the sensitivity of the minimum dispatch time with respect to uncertainty parameters of energy loss and the construction cost with respect to energy dispatch capability is characterized by partitioning the entire space of parameters into multiple subspaces. The efficacy of the developed methodology is demonstrated via case studies with in-depth discussions. Moreover, it is better to investigate a studied energy dispatch problem under uncertainties, such as energy loss percentage and energy supply to the system. Thus, the optimal dispatch time can be considered as a function of these uncertainties. The sensitivity of the minimum dispatch time with respect to uncertainty parameters is characterized by partitioning the entire space of uncertainty parameters into multiple subspaces. They provide decision makers a reference manual for the easy lookup of the possible impact from uncertainties and thus help conduct right decisions. Based on such information, the dynamic impact of different energy dispatch strategies can be evaluated, which will help decision makings under uncertainties. The developed methodology is the first prototype in the field of energy dispatch optimization under emergencies. Although it is based on some simplifications and assumptions, it has built a solid foundation for future in-depth study and real applications.

INTRODUCTION

The natural catastrophe such as tsunamis, earth quake and hurricane will always bring a blow to local economy. Among these natural disasters, the tsunami will have the largest potential to destroy the the living environment. The tsunatami is a series of water waves caused by the displacement of a large volume of a body of water, typically an ocean or a large lake. Earthquakes, volcanic eruptions and other underwater explosions, landslides, glacier calvings, meteorite impacts and other disturbances above or below water all have the potential to generate a tsunami[1]. If the disturbance is close to the coastline, local tsunamis can demolish coastal communities within minutes. A very large disturbance can cause local devastation and export tsunami destruction thousands of miles away. Tsunamis rank high on the scale of natural disasters. Since 1850 alone, tsunamis have been responsible for the loss of over 420,000 lives and billions of dollars of damage to coastal structures and habitats [2].

Among these potential infrastructure damage, the energy shortage is one of the most serious damages in the disaster, especially for tsunatami, because it may cause local areas to suffer from delayed rescues, widespread power outages, tremendous economic losses, and even public safety threats. The vulnerable areas have covered the whole scale of energy system:energy source material, energy storage facilities, energy generation facilities, energy transportation facilities, energy conversion facilities and energy transmission networks. For example,the tsunami that knocked out critical back-up cooling power at the Fukushima Daiichi nuclear power plant on March 11, 2011, is still, in a sense, rolling over Japan's energy system in 2012. Only two of the nation's 54 nuclear power plants are still in operation, and those will be shut down, too, extinguishing the source that provided one-third of the electricity for the world's third-largest national economy before the Tohoku earthquake [3]. The energy shortage will definitely impact the economic development: electricity shortages also increase the threat so that even more manufacturing will flee to the other countries, increasing an exodus well under way before the earthquake. For the nation, an added burden is the increased foreign oil and natural gas dependence to generate power in place of nuclear energy. There has been much public pressure and government commitment to expand renewable energy. Although the country does not have large available land area for massive wind and solar facilities, there have been drives for rooftop solar and research into advancing the technology. Under the consideration of such great impact of the energy for the recovery, the agile energy dispatching through an effective energy transportation network, targeting the minimum energy recovery time, should be a top priority. This is vital to the rebuilding of the other infrastructure systems. The effectiveness and efficiency of agile energy dispatch cannot be achieved only by the human experience because it always calls for a large system of information database and decision making strategy.

Therefore, the application of computer aid decision making will be very essential to handle the above global concern. In this chapter, the introduction and review will be presented for current modeling techniques to provide warning, detection and forecasting for the generated tsunatami. The whole framework will be given for the potential information integration system to handle the emergency planing for the tsunamis and their research progress will be granted.

Predicting when and where the next tsunami will strike is currently impossible. Once the tsunami is generated, forecasting tsunami arrival and impact is possible through modeling and measurement technologies. The current available information system can provide the capability of warning, detection and forecasting for the generated tsunatami[4-6].

Since 1946, the tsunami warning system has provided warnings of potential tsunami danger in the pacific basin by monitoring earthquake activity and the passage of tsunami waves at tide gauges. However, neither seismometers nor coastal tide gauges provide data that allow accurate prediction of the impact of a tsunami at a particular coastal location. Monitoring earthquakes gives a good estimate of the potential for tsunami generation, based on earthquake size and location, but gives no direct information about the tsunami itself. Tide gauges in harbors provide direct measurements of the tsunami, but the tsunami is significantly altered by local bathymetry and harbor shapes, which severely limits their use in forecasting tsunami impact at other locations. Partly because of these data limitations, 15 of 20 tsunami warnings issued since 1946 were considered false alarms because the tsunami that arrived was too weak to cause damage.

Recently developed real-time, deep ocean tsunami detectors will provide the data necessary to make tsunami forecasts. The November 17, 2003, Rat Is. tsunami in Alaska provided the most comprehensive test for the forecast methodology. The Mw 7.8 earthquake on the shelf near Rat Islands, Alaska, generated a tsunami that was detected by three tsunameters located along the Aleutian Trench-the first tsunami detection by the newly developed real-time tsunameter system. These real-time data combined with the model database were then used to produce the real-time model tsunami forecast. For the first time, tsunami model predictions were obtained during the tsunami propagation, before the waves had reached many coastlines. The initial offshore forecast was obtained immediately after preliminary earthquake parameters (location and magnitude Ms = 7.5) became available from the West

Coast/Alaska TWC (about 15-20 minutes after the earthquake). The model estimates provided expected tsunami time series at tsunameter locations. When the closest tsunameter recorded the first tsunami wave, about 80 minutes after the tsunami, the model predictions were compared with the deep-ocean data and the updated forecast was adjusted immediately.

These offshore model scenarios were then used as input for the high-resolution inundation model for Hilo Bay. The model computed tsunami dynamics on several nested grids, with the highest spatial resolution of 30 meters inside the Hilo Bay. None of the tsunamis produced inundation at Hilo, but all of them recorded nearly half a meter (peak-to-trough) signal at Hilo gage. Model forecast predictions for this tide gage are compared with observed data. The comparison demonstrates that amplitudes, arrival time and periods of several first waves of the tsunami wave train were correctly forecasted. More tests are required to ensure that the inundation forecast will work for every likely-to-occur tsunami. When implemented, such forecast will be obtained even faster and would provide enough lead time for potential evacuation or warning cancellation for Hawaii and the U.S. West Coast.

The recent development of real-time deep ocean tsunami detectors and tsunami inundation models has given coastal communities the tools they need to reduce the impact of future tsunamic. The residents who are living in the coastal region will have time to be transferred to a much safer place so that the injuries and casualties while it should be noted that the infrastructure system is difficult to be protected from being destroied even though people can be evacuated. Therefore, the intellgient system integrated the warning system, impact forecasting system and emergency response system will really help to reduce the potential impact of the infrastructure damage and minimize the recovery time.Noted that up to now there is little study in the modeling of emergency planning and infrastructure recovery, our developed mathematical model of economic optimization and dispatch scheduling in this research area will be introduced and reviewed. The developed methodology is the first prototype in the field of energy dispatch optimization under emergencies. Although it is based on some simplifications and assumptions, it has built a solid foundation for future in-depth study and real applications.

Furthermore, our newly developed mathematical models will also be introduced. It is developed for energy network optimization for the consideration of both energy recovery capability and economic factors. The introduced mathematical modeling of economic optimization and dispatch scheduling will provide sufficient technical support to decision-makers. The computer simulation and optimization should provide optimal output solution with the huge input information of variable demand, local energy resources, energy facility capacity, economic sustainability and emergency under extreme conditions. They provide decision makers a reference manual for the easy lookup of the possible impact from uncertainties and thus help conduct right decisions. Based on such information, the dynamic impact of different energy dispatch strategies can be evaluated, which will help decision makings under uncertainties.

CURRENT MODELING TECHNIQUES FOR TSUNAMI

A tsunami is a series of ocean waves generated by sudden displacements in the sea floor, landslides, or volcanic activity. In the deep ocean, the tsunami wave may only be a few inches high. The tsunami wave may come gently ashore or may increase in height to become a fast moving wall of turbulent water several meters high.Predicting when and where the next tsunami will strike is currently impossible. Although a tsunami cannot be prevented, the impact of a tsunami can be mitigated through community preparedness, timely warnings, and effective response. Once the tsunami is generated, forecasting tsunami arrival and impact is possible through modeling and measurement technologies.

Tsunamis are a major concern to the islands and coastal nations because they may occur at any time, with little or no warning, and with destructive force. During the past decade, more than 4,000 people have died from 10 tsunamis in the Pacific. In 1998 alone, the Aitape tsunami killed 2,600 coastal villagers when it struck Papua New Guinea. In Hawaii, more people have died from tsunamis than from hurricanes, earthquakes, and floods combined.

In order to reduce the impact of tsunamis, it is beneficial to have a thorough understanding of the area at risk, its population, infrastructure, and pattern of land use. The objective of tsunami modeling is to develop numerical models of tsunamis propagating through the ocean and striking coastal communities. The theoretical tsunami modeling can help officials understand their risk and vulnerability and assist them in preparing for the possible impact of a major tsunami. Right now there are a lot of models have been developed by the global researh groups and organizations such as Cornell University, Pacific Disaster Center, NOAA Center for Tsunami Research, Centroid-Moment-Tensor Project at Harvard University,United States Geological Survey, Tsunami Computational Portal, Taiwan, International Center for Geohazards, NOAA Pacific Tsunami Warning Center, NOAA West Coast and

Alaska, the World Meteorological Organization/the Intergovernmental Oceanographic Commission , Sea Level Center at University of Hawaii, Permanent Service for Mean Sea Level, National Geophysical Data Center, and Satellite Geodesy at University of California San Diego.

Here, we will give the brief introduction for the COMCOT (Cornell Multi-grid Coupled Tsunami) model, the TTT(the Tsunami Travel Time) model, the MOST (Method of Splitting Tsunami) model, TPC(Tsunami Computational Portal), and GLOSS (The Global Sea Level Observing System).

COMCOT (Cornell Multi-Grid Coupled Tsunami Model) [7]

COMCOT (Cornell Multi-grid Coupled Tsunami Model) is a tsunami modeling package, capable of simulating the entire lifespan of a tsunami, from its generation, propagation and runup/rundown in coastal regions.

Waves can be generated via incident wave maker, fault model, landslide, or even customized profile. Flexible nested grid setup allows for the balance between accuracy and efficiency.

The model has been used to investigate several historical tsunami events, such as the 1960 Chilean tsunami, the 1992 Flores Islands (Indonesia) tsunami [8-9], the 2003 Algeria Tsunami [10] and more recently the 2004 Indian Ocean tsunami [11].

TTT (The Tsunami Travel Time Model) [12]

The Tsunami Travel Time (TTT) model used by Pacific Disaster Center estimates tsunami travel time to all points in the Pacific Basin using a grid format. The TTT model creates a map showing tsunami travel times, as well as a table depicting the travel time as a function of tide gauge locations. When a Tsunami Warning is issued by the Pacific Tsunami Warning Center, the TTT model is activated. The TTT model then creates the travel time map, which is posted to PDC's private website [12].

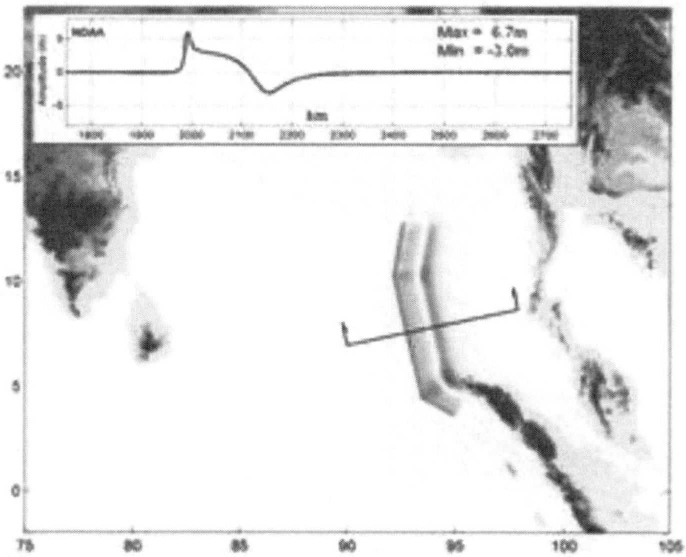

Figure 1. Tsunami Generation.

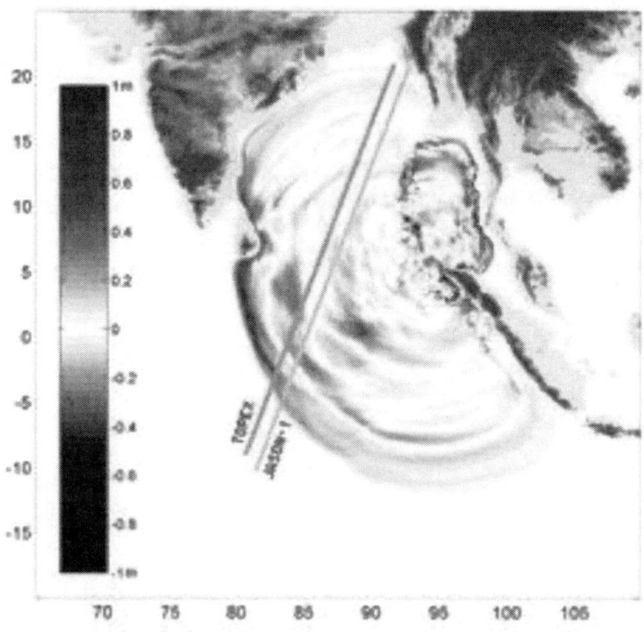

Figure 2. Tsunami Propagation.

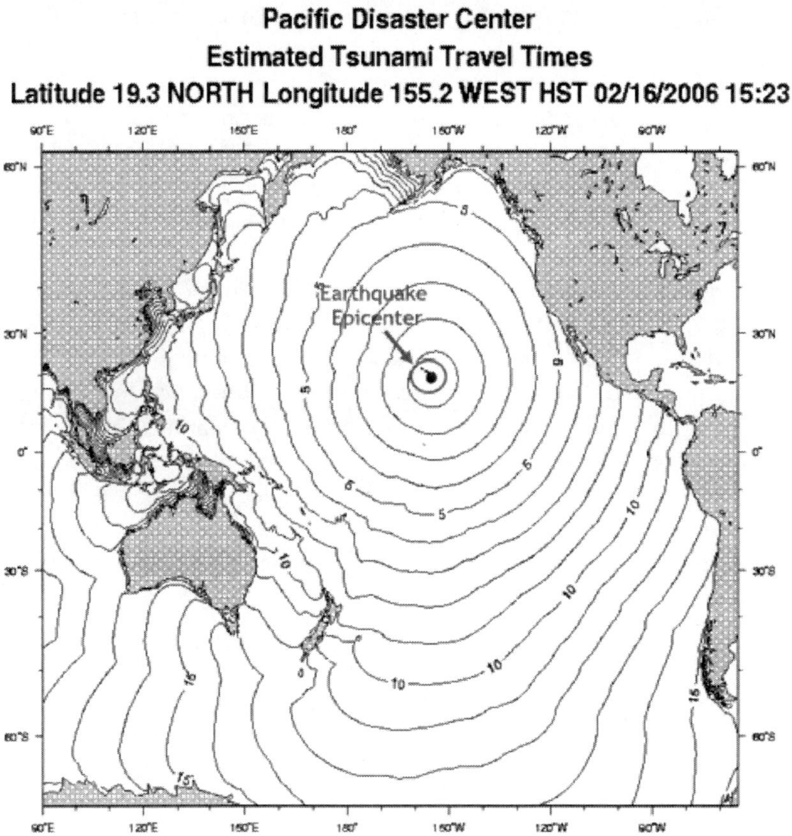

Figure 3. PDC's Tsunami Travel Times map generated on February 16, 2006 3:23 p.m. HST.

MOST (Method of Splitting Tsunami Model) [13]

The MOST (Method of Splitting Tsunami) model, developed by Titov of PMEL and Synolakis of University of Southern California, is the standard model used at the NOAA Center for Tsunami Research[13]. MOST is a suite of numerical simulation codes capable of simulating three processes of tsunami evolution: earthquake, transoceanic propagation, and inundation of dry land. The MOST model has been extensively tested against a number of laboratory experiments and was successfully used for simulations of many historical tsunamis. Several research groups around the world now use MOST for tsunami mitigation.

The MOST has the capability of both forcasting and inundation.The main objective of a forecast model is to provide an estimate of wave arrival time, wave height and inundation area immediately after a tsunami event. Tsunami forecast models are run in real time while a tsunami is propagating in the open ocean, consequently they are designed to perform under very stringent time limitations.

Given the time constraints of this type of study, the process of computing the three stages of tsunami modeling, namely, wave generation, propagation and inundation has been expedited by generating a database of pre-computed scenarios. The pre-computed database contains information about tsunami propagation in the open ocean from a multitude of potential sources. When a tsunami event occurs, an initial source is selected from the pre-computed database. In the initial stages of the tsunami, this selection is based only on the available seismic information for the earthquake event. As the wave propagates across the ocean and successively reaches the DART systems these report the recorded sea level information back to the TWCs which, in turn, process the information and produce a new and more refined estimate of the tsunami source. The result is an increasingly accurate forecast of the tsunami that can be used to issue, watches, warnings or evacuations.

When an event similar to one of the pre-computed scenarios occurs, the available propagation information is used to compute the last stage of the study, wave inundation.

MOST can also provide inundation modeling. An inundation modeling study attempts to recreate the tsunami generation in deep or coastal waters, wave propagation to the impact zone and inundation along the study area. To reproduce the correct wave dynamics during the inundation computations high resolution bathymetric and topographic grids are used in this type of study. The high quality bathymetric and topographic data sets needed for development of inundation maps require maintenance and upgrades as better data becomes available and coastal changes occur.

Inundation studies can be conducted taking a probabilistic approach in which multiple tsunami scenarios are considered, and an assessment of the vulnerability of the coast to tsunami hazard is evaluated, or they may focus on the effect of a particular 'worst case scenario" and assess the impact of such a particularly high impact event on the areas under investigation.

The results of a tsunami inundation study should include information about the maximum wave height and maximum current speed as a function of location, maximum inundation line, as well as time series of wave height at different locations indicating wave arrival time. This information can be used by emergency managers and urban planners primarily to establish evacuation routes and location of vital infrastructure.

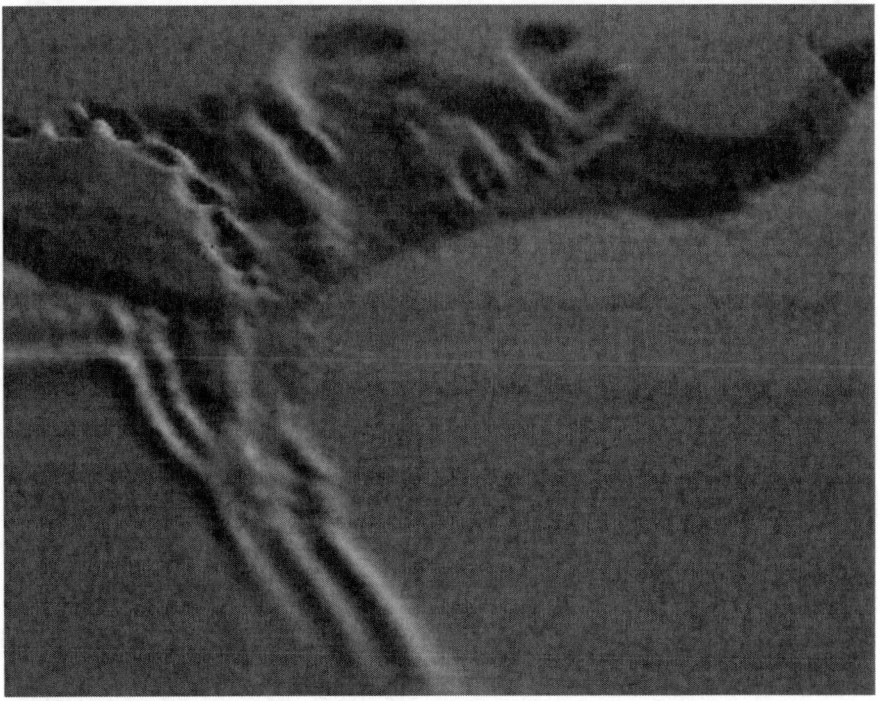

Figure 4. Inundation computed with the MOST model.

In order to predict tsunami creation and inundation, PDC also uses the Method of Splitting Tsunami (MOST) finite difference model, which simulates the formation of shallow water waves. The MOST model takes into account Earth's curvature to predict tsunami travel over long open ocean distances. It also accurately predicts how far the tsunami will travel inland above sea level. The nonlinear equations of the MOST model simulate complex wave dynamics near-shore including tsunami-induced strong currents and wave breaking. PDC implemented the MOST model when looking at the history of tsunamis in Fiji and Vanuatu in order to predict the destruction of future earthquakes.

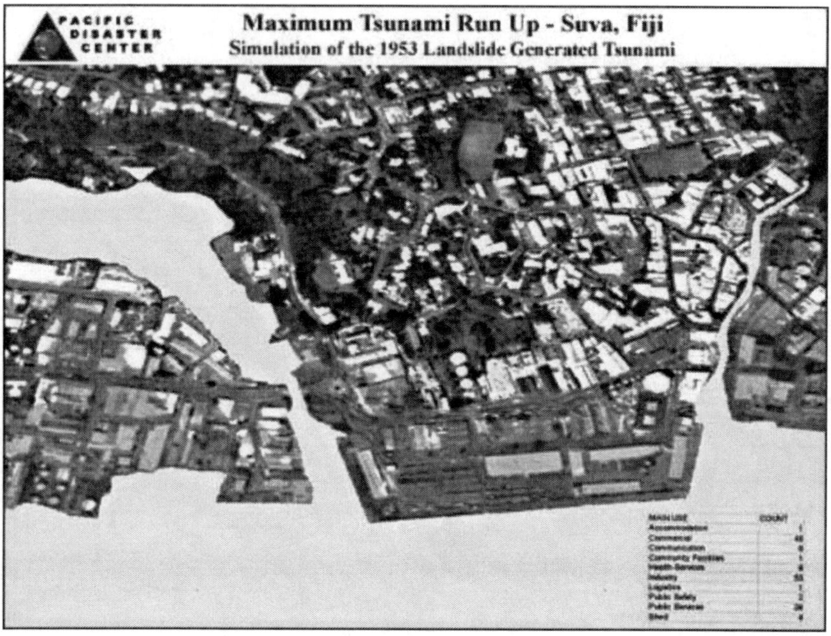

Figure 5. Simulation of the Landslide Generated Tsunami.

TPC (Tsunami Computational Portal) [14]

High-quality tsunami simulation and prediction models exist, but are hard to locate and understand. They are in the hands of individual researchers located at various institutions throughout the world. To obtain critical predictive information (e.g., wave height and propagation estimates, run up effects), each model must be run with particular input data (specific coastline and underwater geography, historical water levels, tidal patterns, etc.). While much of the data is available online, it must be located and gathered from a variety of agencies, then converted to the appropriate formats for each model.

Further, the computing resources for executing the models are scattered, with computer architecture, system availability, and access varying from one location to another. Expert computing knowledge is required to installthe models, expert geographic information knowledge is required to convert and properly align the input data, and expert tsunami knowledge is required to accurately interpret simulation results. The Tsunami Computational Portal

(TCP) provides a collaborative forum where these areas of expertise come together in support of tsunami research and mitigation.

The TCP is being developed jointly by two institutions:University of Alaska Fairbanks(Arctic Region Supercomputing Center and Geophysical Institute) and Oregon State University(Northwest Alliance for Computational Science and Engineering and Department of Civil, Construction & Environmental Engineering with contributions of models from the greater community). Development of the TCP is funded by the Federal government through NOAA.

The goal of conducting tsunami research is to save lives and reduce economic losses due to tsunamis. This requires that we develop sufficient scientific knowledge and appropriate engineering tools on which to base comprehensive tsunami mitigation plans. For a tsunami generated across the ocean, a reliable warning system that predicts the arrival time and expected inundation area can save many lives. It will also reduce economic losses by reducing the numbers of false tsunami evacuation warnings. For a tsunami generated nearby, emergency evacuation must be initiated as soon as the first ground shaking is felt; this will be successful only if an appropriate response plan is already in place and well understood by local inhabitants. The highest priority for mitigating tsunami hazard, therefore, is to identify high tsunami-risk zones and educate citizens who live in and near those zones about proper evacuation procedures.

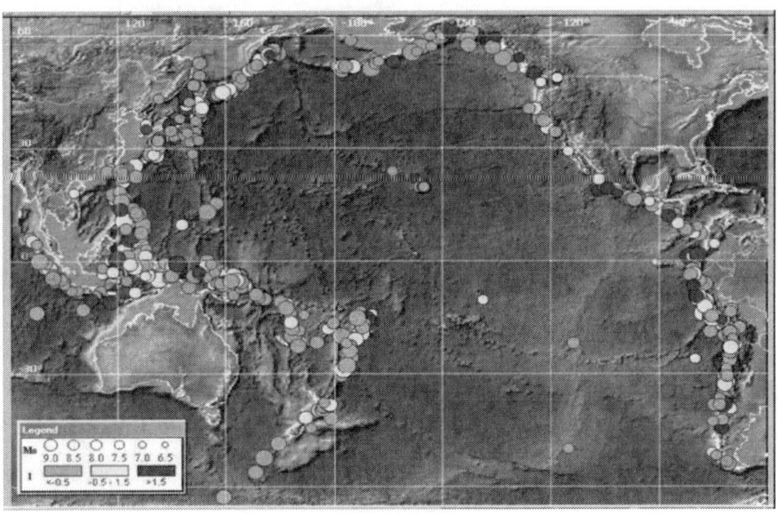

Figure 6. Simulation by Tsunami Computational Portal.

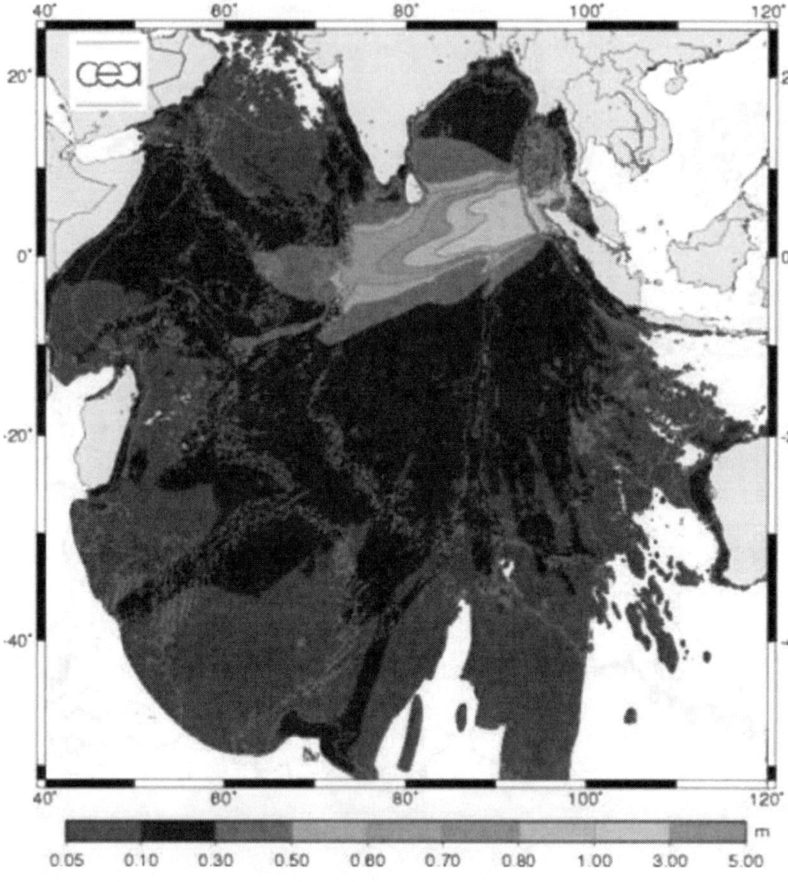

Figure 7. Simulation by Tsunami Computational Portal.

The Tsunami Computational Portal will provide a unique forum to expedite the development of new and enhanced methods for predicting tsunamis and mitigating their effects. Using the portal, researchers and tsunami warning operations staff will be able to collaboratively execute and analyze model behavior, comparing them to the observed effects of past tsunamis. Researchers who have developed computational models "contribute" them to the portal, where they are fully documented and made available for peer review by other tsunami experts. Portal users can select individual models and apply them to a variety of "scenarios," or collections of geographical and infrastructure data that mimic a variety of real-world coastal settings. The models are maintained at a supercomputer center (where they are kept up-to-date by IT professionals) and are executed on behalf of portal users; after execution, users can view or download their results. Most importantly, users can compare the results of running different models on the same "scenarios" and can exchange comments about issues and recommendations with the rest of the tsunami research communit`y. This collaborative process of review and analysis will improve both our understanding of tsunami dynamics and the accuracy of tsunami models.

GLOSS (The Global Sea Level Observing System) [15]

The Global Sea Level Observing System (GLOSS) is an international programme conducted under the auspices of the Joint Technical Commission for Oceanography and Marine Meteorology (JCOMM) of the World Meteorological Organisation (WMO) and the Intergovernmental Oceanographic Commission (IOC). GLOSS aims at the establishment of high quality global and regional sea level networks for application to climate, oceanographic and coastal sea level research. The programme became known as GLOSS as it provides data for deriving the 'Global Level of the Sea Surface'.

The main component of GLOSS is the 'Global Core Network' (GCN) of 290 sea level stations around the world for long term climate change and oceanographic sea level monitoring. The present definition of the GCN (the definition is modified every few years) is called GLOSS10.

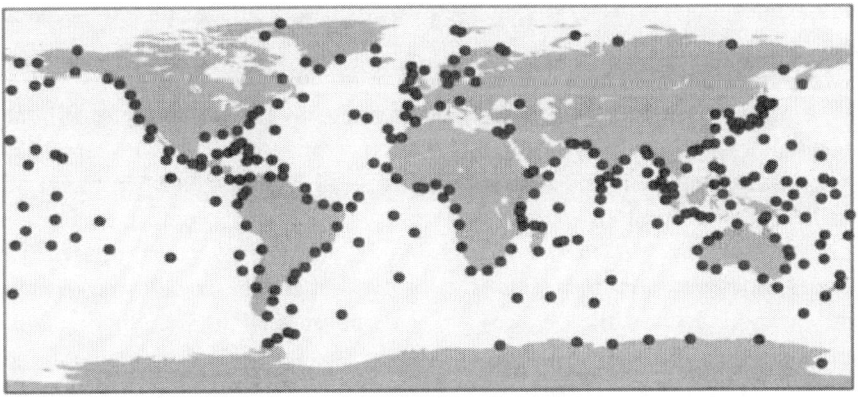

Figure 8. GLOSS Core Network defined by GLOSS10.

The Core Network is designed to provide an approximately evenly-distributed sampling of global coastal sea level variations. Another component is the GLOSS Long Term Trends (LTT) set of gauge sites (some, but not all, of which are in the GCN) for monitoring long term trends and accelerations in global sea level. These will be priority sites for Global Positioning System(GPS) receiver installations to monitor vertical land movements, and their data will contribute to long term climate change studies such as those of the WMO-UNEP Intergovernmental Panel on Climate Change (IPCC).

Motivation of Economic Analysis and Energy Dispatch Optimization for Tsunamis

In recent years, the increasingly frequent occurrences of extreme Tsunami events, such as Japan earthquake and tsunami (2011) have rang the world an alarm on reexamining whether current rescue policies are effective or efficient enough for a speedy recovery of the suffered areas. The tsunami events for the time period from 2010 to 2013 have been listed in below summary tables. These disasters have caused tremendous damages to local areas in terms of economy, people's health and safety, transportation, as well as the energy and the living infrastructures. To restore the functionality and capability of these damaged systems, the recovery of energy system is one of the most important because all the other recovery operations have to be supported by enough available energy. On the other hand, if the local energy shortage caused by a disaster cannot be effectively restored, the local areas will be at risks of rescue delay, widespread power outages, economic losses, and even public safety threats. Therefore, the recovery time minimization of a suffered energy network system should be a top priority of an emergency rescue plan.

Nowadays energy supply is characterized by its diversity, including traditional energy such as fossil fuels, nuclear power, as well as renewable energy such as solar, hydroelectric, geothermal, biomass, and wind energy. It also involves a complex network system composed of energy generation, energy transformation, energy transportation, and energy consumption. The network does provide the great flexibility for energy transformation and transportation; meanwhile, it presents a complex task for conducting agile energy dispatching when extreme events have caused local energy shortages that need to be restored timely. Conceivably, any type of dispatched energy under certain emergency condition has its own characteristics in terms of availability, quantity, transportation speed, and conversion rate and efficiency to other types of energy. Thus, different types of energy should be dispatched through a superior plan. For instance, energy sources such as petroleum or coal can be directly transported to a suffered area; meanwhile, they can also be converted to electricity in a source region and then sent to the suffered area through an available electricity network. Sometimes, even the transportation of the same type of energy may have different alternative routes for selection, which needs to be optimally determined from the view point of the entire energy dispatch system.

Facing the challenges of emergency response to energy shortage, decision makers often encounters various uncertainties that inevitably influence the performance of a being designated energy dispatch plan. The uncertainties can upset the optimality and even the feasibility of the dispatch plan. Thus, quantitative analysis on the impact of uncertainties is of great significance for the study of energy network dispatch under emergencies. Technically, a viable approach is to conduct a full evaluation of the effects of uncertainties based on all their possibilities. This will provide decision makers a complete roadmap of the space of uncertainty parameters. Through this way, the objective function and the optimization parameters are represented as functions of uncertainty parameters (i.e., parametric programming); meanwhile, the regions in the space of the uncertainties characterized by these functions can be obtained.

From the literature survey, the reported work on local energy shortage and management problems has been studied. The contribution includes policy making of normal management [16], energy operation in buildings [17], economic impact from energy shortage [18], energy shortage solution by economic lever function [19], typical technology proposal of renewable energy utility [20], and energy system optimization for reduction of environmental impact [21]. However, there is still lack of study on decision marking support for energy dispatch optimization under emergency situations. Conceivably, this is not a trivial task, because it involves multiple types of energy (traditional energy and renewable energy), different energy shortage modes (e.g., short of energy source materials or incapable of energy transformation or transportation), and various energy dispatch scenarios (e.g., energy dispatch from one place to another place, 1→1for short; or possibly 1→n, n→1, and n→n) based on the available infrastructure of local energy dispatch network.

In the field of parametric programming for handling uncertainties, Gal and Nedoma [22] developed an algorithm to solve multi-parametric linear programming problems based upon the Simplex algorithm. Acevedo and Pistikopoulos [23] extended the field of parametric programming to address multi-parametric mixed-integer linear programs, and Dua, Bozinis and Pistikopoulos [24] made significant contributions in the area of multi-parametric mixed-integer quadratic problems. Acevedo and Salgueiro [25] presented a novel algorithm for the solution of convex multi-parametric nonlinear programming problems. More studies have also been reported in this area from Acevedo and Pistikopoulos [26] and [27]; Ryu, Dua, and Pistikopoulos [28]; Jia and Ierapetritou, [29] and [30]; Verderame and Floudas [31].

Our research is aimed to development a novel methodology for energy network dispatch optimization under emergency of local energy shortage. It includes four stages of work. First, emergency-area-centered energy network needs to be characterized, where the capacity, quantity, and availability of various energy sources are determined. Second, the energy initial situation under emergency conditions needs to be identified. Then, the energy dispatch optimization is conducted based on a developed MILP (mixed-integer linear programming) model in the third stage. Finally, the sensitivity of minimum dispatch time with respect to uncertainty parameters is characterized by partitioning the entire space of uncertainty parameters into multiple subspaces. The developed methodology is the first prototype in the field of energy dispatch optimization under emergencies. Although it is based on some simplifications and assumptions, it has built a solid foundation for future in-depth study and real applications.

Motivation of Economic Analysis and Energy Dispatch Optimization for Tsunamis

Since it is only the priliminary study in this area, our research considers a region containing multiple living areas (e.g., towns or cities). To support these areas, energy production and consumption must exist in this region. This can be considered as an energy distribution and exchange network, where each living area is considered as a node having its own energy generation, transformation, consumption, and storage system; meanwhile, multiple energy dispatch infrastructures are available, such as highway, railway, power line, and petroleum pipelines, which enable the distribution of multiple energy sources among those energy nodes. The energy transportations between two nodes are also called connections. Note that in such an energy dispatch network, each node has limited capacities of storage for different types of energy source, energy generation, and energy transformation; meanwhile, each energy transportation connection also has a restrained capability.

Under extreme conditions of natural disaster and accidents, both energy nodes and connections of a given energy dispatch network may suffer from potential damages. The damages may be reflected as the breakdown of transportation pipeline, losses of energy source materials, or failure of energy transformation equipment (e.g., the damage of power station can fully or partially disable the electricity generation from coal or petroleum fuels). In such a damaged energy network, some nodes may suddenly be in the trouble of energy shortage. To recover the energy supply as soon as possible, the entire region should be motivated to support the energy shortage areas, i.e., through available infrastructure, dispatching various energy or energy source materials from other areas to the in need areas timely. Note that different energy sources have different dispatch ways and speeds based on an available infrastructure. To minimize the energy recovery time, optimal decisions will be made to determine how to select appropriate energy sources and conduct smart dispatching methods, which involves an optimization problem. For clarity, the optimal energy dispatch problem can be summarized as below.

Assumptions:

1) After extreme events, the energy demand and current inventory for any type of energy in each node are predictable.
2) The energy type does not change during the transportation; energy can only be transformed to another type at a node;
3) After extreme events, the capacity for different energy transformation at each node and capability for energy transportation among different nodes based on available infrastructure are predicable.
4) During the time period of energy dispatch, there will be no further damage to the available energy transformation and transportation capability.
5) For any type of energy, the dispatch speed from or to an energy node is considered as constant.

Given information:

1) Energy demand and inventory for any type of energy at each node;

2) The capability and efficiency of energy transformation in each node after extreme events;
3) The connection availability of the energy dispatch network after extreme events, which is used to determine the energy transportation route and speed.

Information to be determined:

1) The minimum recovery time for energy network restoration;
2) The energy dispatch strategy, including energy transformation plan in each node and energy transportation plan among different nodes;
3) Partition of optimal energy dispatch strategies regarding the uncertainty of energy shortage conditions.

METHODOLOGY FRAMEWORK

This methodology has included four stages as shown in Figure 9. In the first stage, the energy network for emergency responsible area is characterized. The capacity, quantity, availability, and the convertibility of various energy sources in each energy node will be determined. The considered energy sources may include petroleum, nuclear electric power, natural gas, coal, and the renewable energy (solar energy, geothermal energy, biomass, and wind energy). In the second stage, under emergency of local energy shortage, the energy initial condition for each node of the studied energy network will be evaluated and estimated. The energy initial condition is generally an update of energy scenarios from the first stage, affected by the emergency events case by case. Energy shortage information should be collected in this stage.

Based on the estimated energy initial conditions, in the third stage energy dispatch optimization based on a developed MILP model will be conducted, which will integrate information of input and output of energy resource, energy demand, inventory, and transformation and transportation methods to minimize the recovery time for local energy restoration. It is a general energy dispatch model, which can cover all the energy dispatch scenarios (e.g., 1→1, 1→n, n→1, or n→n). Available commercial solver, such as CPLEX [32] and [33], can be employed to solve the MILP model and obtain the global optimal solution. Since the optimal energy dispatch solution is obtained based on the damage evaluations of the energy network. Uncertainties, such as energy loss percentage and energy supply to the network system, have to be addressed during the decision making. They can affect the optimality and even feasibility of the obtained dispatch solution. Thus, the sensitivity of minimum dispatch time to uncertainty parameters is characterized by partitioning the entire space of uncertainty parameters into multiple subspaces in the fourth stage. For each of the partitioned subspaces, the minimum dispatch time with respect to the uncertainty parameters within the subspace has the same representation function. It actually discloses a roadmap of uncertainty impacts to energy dispatch solutions, based on which in-depth analysis can be conducted to support energy dispatch decisions.

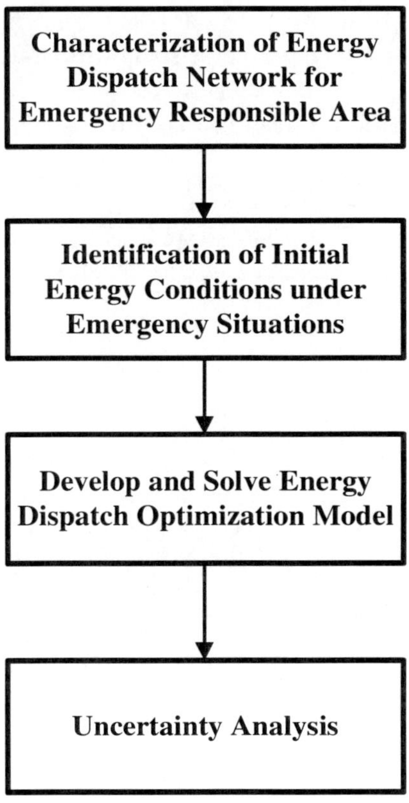

Figure 9. General methodology framework.

APPLICATION DEMONSTRATION

The developed methodology for energy network dispatch optimization under emergency of local energy shortage includes four stages of work. First, emergency-area-centered energy network needs to be characterized, where the capacity, quantity, and availability of various energy sources are determined. Second, the energy initial situation under emergency conditions needs to be identified. Then, the energy dispatch optimization is conducted based on a developed MILP (mixed-integer linear programming) model in the third stage. Finally, the sensitivity of the minimum dispatch time with respect to uncertainty parameters is characterized by partitioning the entire space of uncertainty parameters into multiple subspaces. The efficacy of the developed methodology is demonstrated via a case study with in-depth discussions.

Table 1. Given Data of the Case Study

Description	Parameter	Energy Source	Unit	i/j		
				A/B or C	**B/A or C**	**C/A or B**
Electricity Conversion Efficiency from Other Energy Sources (k: electricity)	$\eta_{i,l,k}$	l: Petroleum Oil	MWh/Barrel	0.55	0.51	0.62
		l: Natural Gas	MWh/Million ft^3	0.12	0.125	0.11
		l: Coal	MWh/Short Ton	1.5	1.2	1.6
Hourly Electricity Generation Limit from Source Materials (k: electricity)	$EG^u_{i,l,k}$	l: Petroleum Oil	MWh/hr	75	65	90
		l: Natural Gas	MWh/hr	500	400	600
		l: Coal	MWh/hr	50	40	45
Hourly Limit of Transportation Input from City j to City i	$EI^u_{j,i,k}$	k: Petroleum Oil	Barrel/hr	0.03	0.03	0.03
		k: Natural Gas	Million ft^3/hr	0.2	0.2	0.2
		k: Coal	Short Ton/hr	0.01	0.01	0.01
		k: Electricity	MWh/hr	0.05	0.05	0.05
Hourly Limit of Transportation Output from City i to City j	$EO^u_{i,j,k}$	k: Petroleum Oil	Barrel/hr	0.03	0.03	0.03
		k: Natural Gas	Million ft^3/hr	0.2	0.2	0.2
		k: Coal	Short Ton/hr	0.01	0.01	0.01
		k: Electricity	MWh/hr	0.05	0.05	0.05
Required Energy Inventory Amount for City i	$RE_{i,k}$	k: Petroleum Oil	Barrel	400	426	493
		k: Natural Gas	Million ft^3	500	700	600
		k: Coal	Short Ton	314	346	396
		k: Electricity	MWh	1600	1550	1600
Hourly Energy Supply for City i	$S_{i,k,t}$	k: Petroleum Oil	Barrel/hr	27.45	29.25	27
		k: Natural Gas	Million ft^3/hr	135	180	225
		k: Coal	Short Ton/hr	27	22.5	27
		k: Electricity	MWh/hr	0	0	0

To demonstrate the efficacy of the developed methodology, a case study has been conducted and discussed. An energy network composed of three cities (A, B, and C) was influenced by an extreme weather event, the infrastructure of city A got partially damaged and is currently experiencing energy shortage. The other two cities (B and C) are not affected by the event and thus are able to provide energy and source materials to support city A. By following the first two stages of the proposed methodology, the energy initial conditions of this case study are summarized in Table 1, which gives all the parameters used in the energy dispatch model. Beside the listed parameters, the efficiency of electricity transportation is approximated as 0.95, suggesting 5% loss during electricity network transmission; the hourly energy consuming index is 0.05, suggesting 5% of the available inventory energy will be consumed from the network. Meanwhile, there is no time retention for electricity transmission between cities while the transportation time for source material is two hours between two different cities.

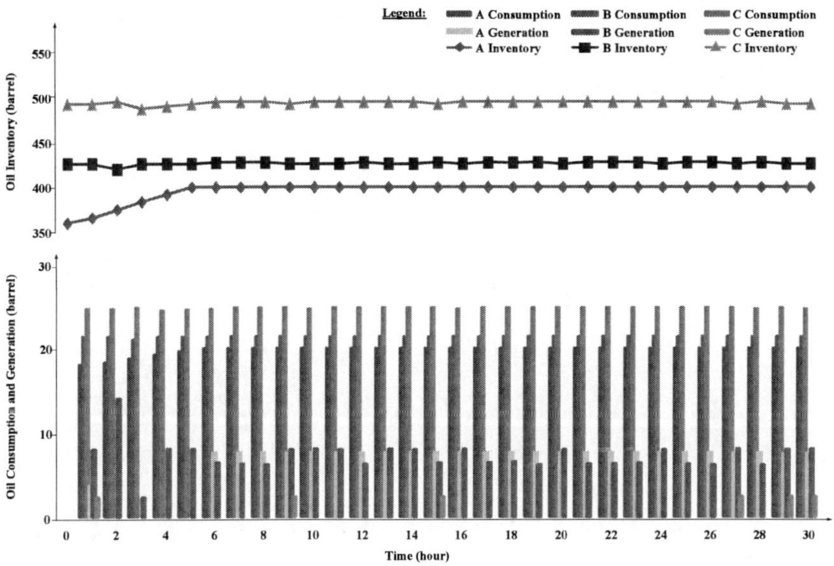

Figure 10. Dynamic profiles of petroleum oil inventory and consumption in different cities.

During the emergency situation, assume all types of energy suffer 10% loss in city A, meanwhile there is no energy loss damage to the other cities. Note that the optimization is performed to deal with an MILP problem.

Because the number of variables and constraints involved is not large, the global solution optimality can be guaranteed by employing the commercial solvers such as CPLEX used in this study. Based on the developed energy dispatch model with GAMS [18], the optimization result shows that the minimum dispatch time for this energy restoration of city A is 26 hours. The dynamic profiles for the energy dispatch within the network are presented in Figures 10 through 18.

Figures 10 through 13 show the dynamic energy inventory (represented by bar chart) and energy consumption (represented by trend curves) of petroleum oil (see Figure 10), coal (Figure 11), natural gas (Figure 12), and electricity (Figure 13) in different cities during the energy dispatch period. Figures 10 through 12 indicate that the energy sources of petroleum oil, coal, and natural gas in city A have reached the required quantity within 6 hours. However, electricity recovery picks up slowly because: i) during the first 6 hours, the electricity source materials are insufficient and unstable to fully support electricity generation; and ii) three cities all have small electricity generation capacity, even surplus electricity generated from cities B and C are fully transmitted to city A already. Figure 13 shows that the electricity restoration time in city A is 26 hours.

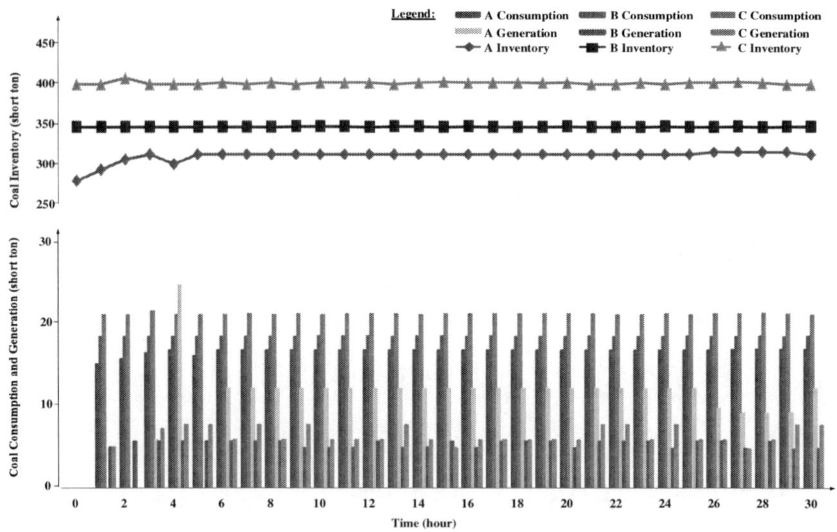

Figure 11. Dynamic profiles of coal inventory and consumption in different cities.

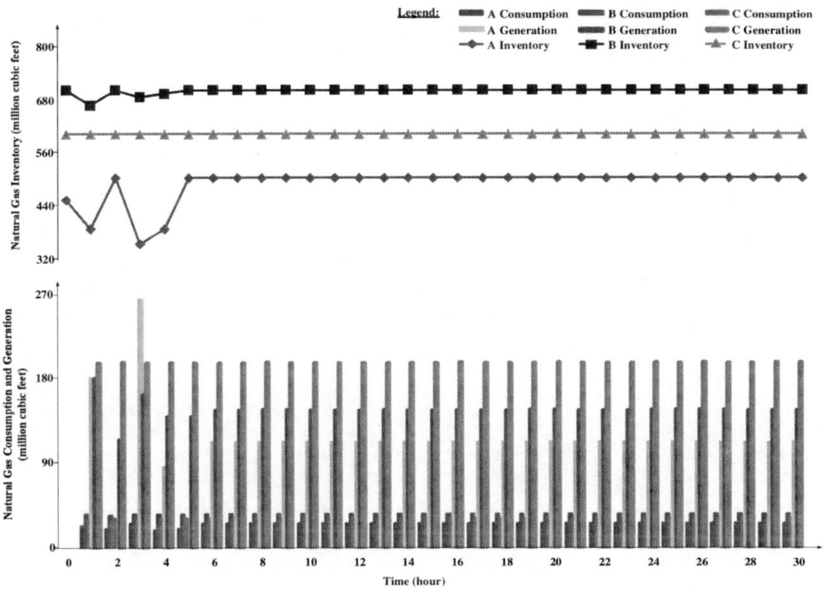

Figure 12. Dynamic profiles of natural gas inventory and consumption in different cities.

From Figures 10 through 13, energy profiles in cities B and C only experience small upsets during the energy dispatch period; while those of city A change a lot because it suffers from energy shortage. Especially from Figures 11 through 13, the energy source consumption for electricity generation in city A changes dramatically. This is because city A needs to use the most efficient and available energy sources to generate electricity, so as to accomplish electricity restoration as soon as possible. Thus, in the first two hours city A is with insufficient energy sources for electricity generation. City A uses large quantity of natural gas for electricity generation in the 3rd hour (see Figure 12) because of the high conversion efficiency; and thus the natural gas inventory has decreased a lot in the 3rd hour. After that, because the natural gas inventory needs to pick up, petroleum oil and coal consumptions for electricity generation in city A increase (see Figures 10 and 11). Starting from the 6th hour, all the energy sources are consumed in full capacity for electricity generation in city A.

Figure 14 gives a 3D view of the accumulative transportation amount of various energy resources between any pair of cities. The three coordinates represent the accumulative transportation amounts of petroleum oil, coal, and natural gas, respectively. As time elapses, the accumulative transportation amount is always non-decreasing. Thus, arrows of each curve indicate direction of change with respect to dispatching time. It shows that the accumulative transportation amounts of various energy resources from city A to city B and from city A to city C keep zero until the inventory quantity in city A has been satisfied; while the transportation amounts from cities B and C to A are most significant, although the mutual transportations between B and C are also nontrivial. Furthermore, Figure 14 shows that the hourly transportation amounts from both cities B and C to A are in their full capacities until the near end of the time period. Therefore, both accumulative transportation increments are changing linearly in a wide range in the 3D plot.

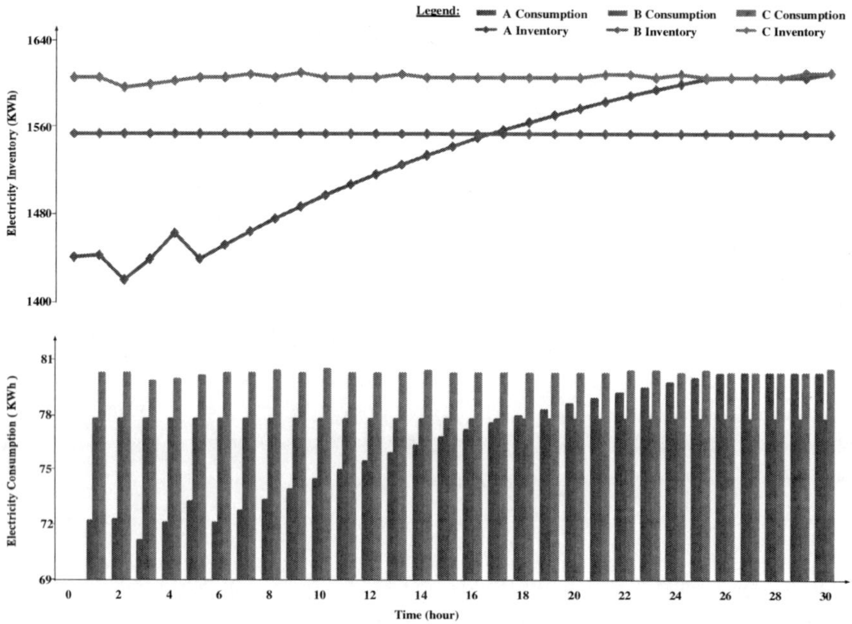

Figure 13. Dynamic profiles of electricity inventory and consumption in different cities.

To disclose more details of energy source transportation between each pair of cities, Figures 15 through 18 show accumulative transportation amounts of the petroleum oil, natural gas, coal, and electricity, respectively. Because the energy source materials dispatched from one city to another will take two hours to reach the destination, the curves in Figures 15 through 17 (respectively addressing petroleum oil, natural gas, and coal) clearly show such a transportation time delay. Note that the time delay during electricity transportation (see Figure 18) between two cities has been ignored; however, it costs 5% loss during the transportation. Thus, the electricity arrival and departure curves between two cities are always in 95% proportion in Figure 18. Figures 15 through 18 also show no energy output from city A because of its energy shortage. Cities B and C are dispatching the energy material with the highest loading to city A, which is consistent with the previous analysis based on Figure 14. After the 26th hour when city A has restored all energy, the energy dispatch activity from the other two cities reduces.

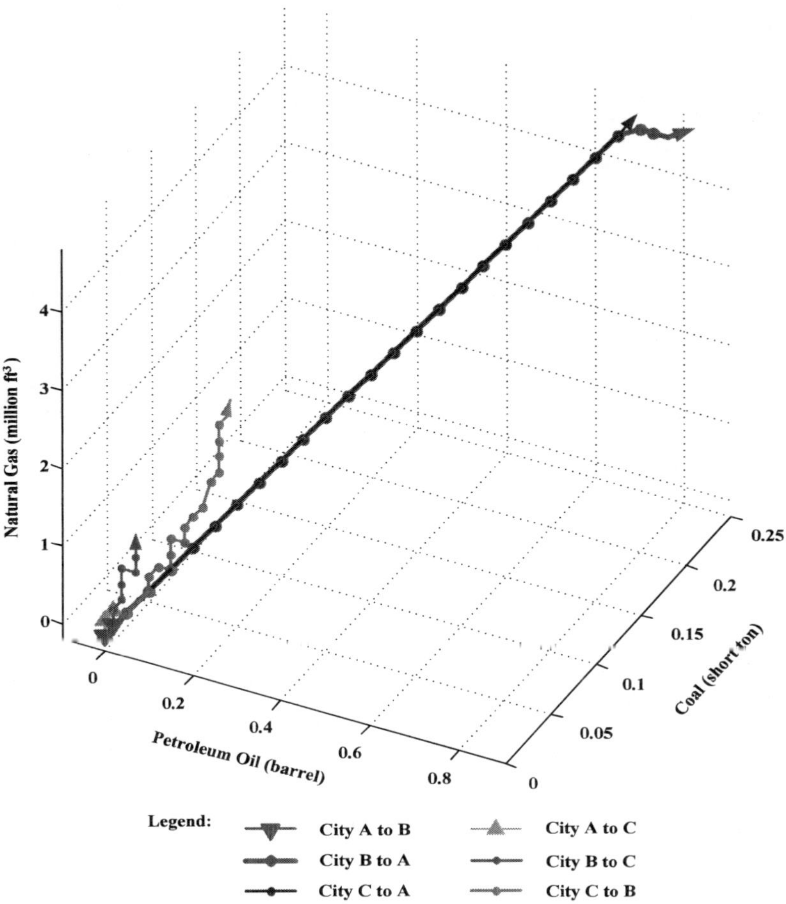

Figure 14. Accumulative transportation amount of various energy resources between cities.

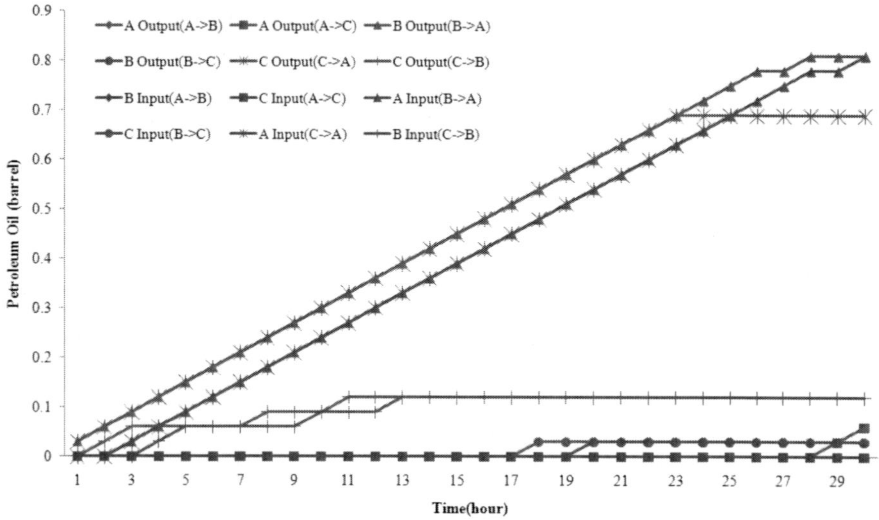

Figure 15. Accumulative transportation amount of petroleum oil between cities.

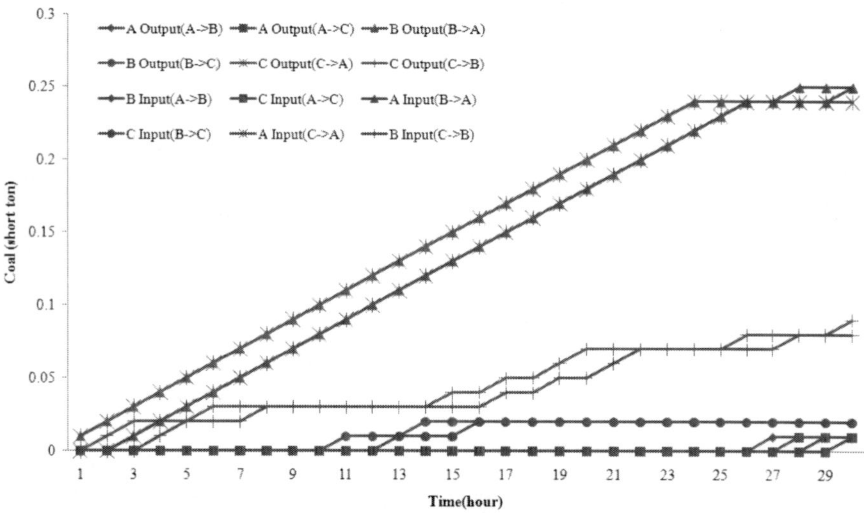

Figure 16. Accumulative transportation amount of coal between cities.

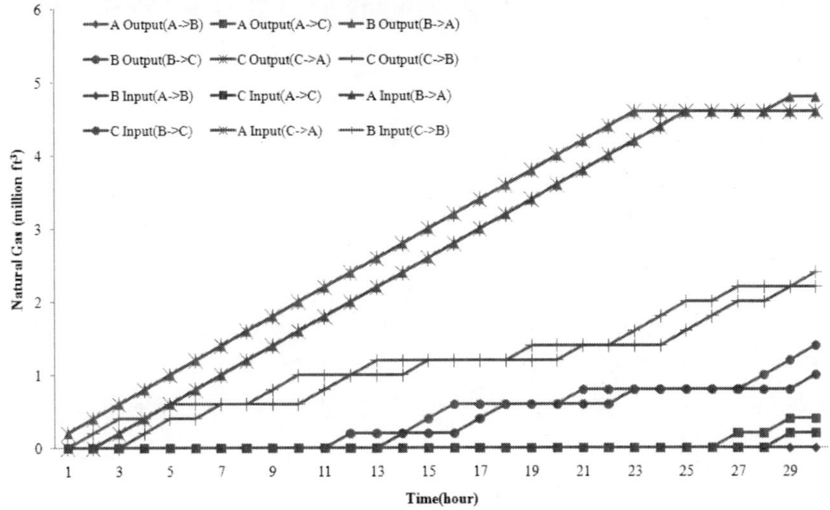

Figure 17. Accumulative transportation amount of natural gas between cities.

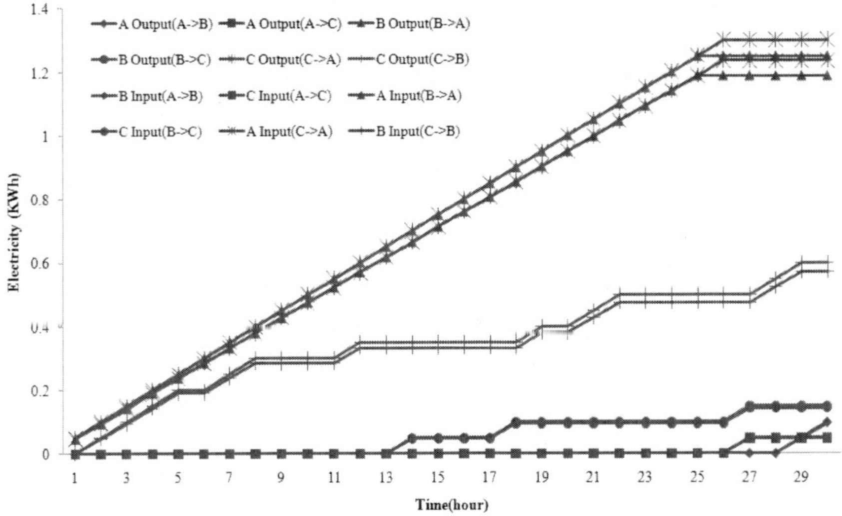

Figure 18. Accumulative transportation amount of electricity between cities.

Table 2. Uncertainty Space Partitioning Results for the Case Study

Region	Vertex	θ1	θ2	M(θ)	A(θ)	Dispatch Time T	AVG θ1	AVG θ2	M(θ)	A(θ)
CR1	P1	0	0	13	13	$T=\frac{160}{3}\theta_1+160\theta_2+13$	0.01875	0.01875	17	17
	P2	0.0375	0	15	15					
	P3	0.0375	0.0375	21	21					
	P4	0	0.0375	19	19					
CR2	P1	0.0375	0	15	15	$T=\frac{160}{3}\theta_1+160\theta_2+13$	0.046875	0.009375	17	17
	P2	0.05625	0	16	16					
	P3	0.05625	0.01875	19	19					
	P4	0.0375	0.01875	18	18					
CR3	P1	0.05625	0	16	16	$T=\frac{160}{3}\theta_1+160\theta_2+13$	0.065625	0.009375	18	18
	P2	0.075	0	17	17					
	P3	0.075	0.01875	20	20					
	P4	0.05625	0.01875	19	19					
CR4	P1	0.05625	0.01875	19	19	$T=\frac{160}{3}\theta_1+\frac{320}{3}\theta_2+14$	0.065625	0.028125	21	21
	P2	0.075	0.01875	20	20					
	P3	0.075	0.0375	22	22					
	P4	0.05625	0.0375	21	21					
CR5	P1	0.0375	0.01875	18	18	$T=\frac{80}{3}\theta_1+\frac{400}{3}\theta_2+14.75$	0.046875	0.028125	20	20
	P2	0.05625	0.01875	19	19					
	P3	0.05625	0.0375	21	21					
	P4	0.0375	0.0375	21	21					
CR6	P1	0.0375	0.0375	21	21	$T=\frac{80}{3}\theta_1+\frac{320}{3}\theta_2+16$	0.05625	0.05625	23	23
	P2	0.075	0.0375	22	22					
	P3	0.075	0.075	26	26					
	P4	0.0375	0.075	25	25					
CR7	P1	0	0.0375	19	19	$T=\frac{80}{3}\theta_1+\frac{400}{3}\theta_2+14.25$	0.009375	0.046875	21	21
	P2	0.01875	0.0375	20	20					
	P3	0.01875	0.05625	22	22					
	P4	0	0.05625	22	22					
CR8	P1	0.01875	0.0375	20	20	$T=\frac{160}{3}\theta_1+\frac{320}{3}\theta_2+15$	0.028125	0.046875	21	21
	P2	0.0375	0.0375	21	21					
	P3	0.0375	0.05625	23	23					
	P4	0.01875	0.05625	22	22					
CR9	P1	0.01875	0.05625	22	22	$T=\frac{160}{3}\theta_1+\frac{320}{3}\theta_2+15$	0.028125	0.065625	24	24
	P2	0.0375	0.05625	23	23					
	P3	0.0375	0.075	25	25					
	P4	0.01875	0.075	24	24					
CR10	P1	0	0.05625	22	22	$T=\frac{320}{3}\theta_2+16$	0.009375	0.065625	23	23
	P2	0.01875	0.05625	22	22					
	P3	0.01875	0.075	24	24					
	P4	0	0.075	24	24					
CR11	P1	0.075	0	17	17	$T=\frac{80}{3}\theta_1+\frac{400}{3}\theta_2+15.25$	0.084375	0.009375	19	19
	P2	0.09375	0	18	18					
	P3	0.09375	0.01875	20	20					
	P4	0.075	0.01875	20	20					
CR12	P1	0.09375	0	18	18	$T=\frac{160}{3}\theta_1+\frac{320}{3}\theta_2+13$	0.103125	0.009375	20	20
	P2	0.1125	0	19	19					
	P3	0.1125	0.01875	21	21					
	P4	0.09375	0.01875	20	20					

Region	Vertex	θ1	θ2	M(θ)	A(θ)	Dispatch Time T	AVG θ1	AVG θ2	M(θ)	A(θ)
CR13	P1	0.09375	0.01875	20	20	$T=\frac{80}{3}\theta_1+\frac{400}{3}\theta_2+15.25$	0.103125	0.028125	22	22
	P2	0.1125	0.01875	21	21					
	P3	0.1125	0.0375	23	23					
	P4	0.09375	0.0375	23	23					
CR14	P1	0.075	0.01875	20	20	$T=\frac{80}{3}\theta_1+\frac{400}{3}\theta_2+15.25$	0.084375	0.028125	21	21
	P2	0.09375	0.01875	20	20					
	P3	0.09375	0.0375	23	23					
	P4	0.075	0.0375	22	22					
CR15	P1	0.1125	0	19	19	$T=\frac{80}{3}\theta_1+\frac{320}{3}\theta_2+16$	0.13125	0.01875	22	22
	P2	0.15	0	20	20					
	P3	0.15	0.0375	24	24					
	P4	0.1125	0.0375	23	23					
CR16	P1	0.1125	0.0375	23	23	$T=\frac{80}{3}\theta_1+\frac{320}{3}\theta_2+16$	0.13125	0.05625	26	26
	P2	0.15	0.0375	24	24					
	P3	0.15	0.075	28	28					
	P4	0.1125	0.075	27	27					
CR17	P1	0.075	0.0375	22	22	$T=\frac{80}{3}\theta_1+\frac{320}{3}\theta_2+16$	0.09375	0.05625	25	25
	P2	0.1125	0.0375	23	23					
	P3	0.1125	0.075	27	27					
	P4	0.075	0.075	26	26					
CR18	P1	0.075	0.075	26	26	$T=20\theta_1+\frac{220}{3}\theta_2+19.25$	0.1125	0.1125	30	30
	P2	0.15	0.075	28	28					
	P3	0.15	0.15	33	33					
	P4	0.075	0.15	32	32					
CR19	P1	0	0.075	24	24	$T=\frac{80}{3}\theta_1+80\theta_2+18$	0.0375	0.1125	28	28
	P2	0.075	0.075	26	26					
	P3	0.075	0.15	32	32					
	P4	0	0.15	30	30					
CR20	P1	0.15	0	20	20	$T=\frac{160}{3}\theta_1+\frac{320}{3}\theta_2+12$	0.159375	0.009375	22	22
	P2	0.16875	0	21	21					
	P3	0.16875	0.01875	23	23					
	P4	0.15	0.01875	22	22					
CR21	P1	0.16875	0	21	21	$T=\frac{80}{3}\theta_1+\frac{400}{3}\theta_2+16.25$	0.178125	0.009375	22	22
	P2	0.1875	0	21	21					
	P3	0.1875	0.01875	24	24					
	P4	0.16875	0.01875	23	23					
CR22	P1	0.16875	0.01875	23	23	$T=\frac{160}{3}\theta_1+\frac{320}{3}\theta_2+12$	0.178125	0.028125	24	24
	P2	0.1875	0.01875	24	24					
	P3	0.1875	0.0375	26	26					
	P4	0.16875	0.0375	25	25					
CR23	P1	0.15	0.01875	22	22	$T=\frac{160}{3}\theta_1+\frac{320}{3}\theta_2+12$	0.159375	0.028125	24	24
	P2	0.16875	0.01875	23	23					
	P3	0.16875	0.0375	25	25					
	P4	0.15	0.0375	24	24					
CR24	P1	0.1875	0	21	21	$T=\frac{80}{3}\theta_1+\frac{400}{3}\theta_2+16.25$	0.196875	0.009375	23	23
	P2	0.20625	0	22	22					
	P3	0.20625	0.01875	24	24					
	P4	0.1875	0.01875	24	24					

Table 2. (Continued)

Region	Vertex	θ1	θ2	M(θ)	A(θ)	Dispatch Time T	AVG θ1	AVG θ2	M(θ)	A(θ)
CR25	P1	0.20625	0	22	22	$T=\frac{160}{3}\theta_1+\frac{320}{3}\theta_2+11$	0.215625	0.009375	23	23
	P2	0.225	0	23	23					
	P3	0.225	0.01875	25	25					
	P4	0.20625	0.01875	24	24					
CR26	P1	0.20625	0.01875	24	24	$T=\frac{160}{3}\theta_1+\frac{320}{3}\theta_2+11$	0.215625	0.028125	25	25
	P2	0.225	0.01875	25	25					
	P3	0.225	0.0375	27	27					
	P4	0.20625	0.0375	26	26					
CR27	P1	0.1875	0.01875	24	24	$T=\frac{320}{3}\theta_2+22$	0.196875	0.028125	25	25
	P2	0.20625	0.01875	24	24					
	P3	0.20625	0.0375	26	26					
	P4	0.1875	0.0375	26	26					
CR28	P1	0.1875	0.0375	26	26	$T=\frac{80}{3}\theta_1+80\theta_2+18$	0.20625	0.05625	28	28
	P2	0.225	0.0375	27	27					
	P3	0.225	0.075	30	30					
	P4	0.1875	0.075	29	29					
CR29	P1	0.15	0.0375	24	24	$T=\frac{160}{3}\theta_1+\frac{320}{3}\theta_2+12$	0.159375	0.046875	26	26
	P2	0.16875	0.0375	25	25					
	P3	0.16875	0.05625	27	27					
	P4	0.15	0.05625	26	26					
CR30	P1	0.16875	0.0375	25	25	$T=\frac{80}{3}\theta_1+80\theta_2+17.75$	0.178125	0.046875	26	26
	P2	0.1875	0.0375	26	26					
	P3	0.1875	0.05625	27	27					
	P4	0.16875	0.05625	27	27					
CR31	P1	0.16875	0.05625	27	27	$T=\frac{80}{3}\theta_1+80\theta_2+17.75$	0.178125	0.065625	28	28
	P2	0.1875	0.05625	27	27					
	P3	0.1875	0.075	29	29					
	P4	0.16875	0.075	28	28					
CR32	P1	0.15	0.05625	26	26	$T=\frac{80}{3}\theta_1+80\theta_2+17.75$	0.159375	0.065625	27	27
	P2	0.16875	0.05625	27	27					
	P3	0.16875	0.075	28	28					
	P4	0.15	0.075	28	28					
CR33	P1	0.225	0	23	23	$T=\frac{80}{3}\theta_1+\frac{320}{3}\theta_2+17$	0.24375	0.01875	25	25
	P2	0.2625	0	24	24					
	P3	0.2625	0.0375	28	28					
	P4	0.225	0.0375	27	27					
CR34	P1	0.2625	0	24	24	$T=\frac{40}{3}\theta_1+\frac{280}{3}\theta_2+20.75$	0.28125	0.01875	26	26
	P2	0.3	0	25	25					
	P3	0.3	0.0375	28	28					
	P4	0.2625	0.0375	28	28					
CR35	P1	0.2625	0.0375	28	28	$T=80\theta_2+25$	0.28125	0.05625	30	30
	P2	0.3	0.0375	28	28					
	P3	0.3	0.075	31	31					
	P4	0.2625	0.075	31	31					

Region	Vertex	θ1	θ2	M(θ)	A(θ)	Dispatch Time T	AVG θ1	AVG θ2	M(θ)	A(θ)
CR36	P1	0.225	0.0375	27	28	$T=\frac{80}{3}\theta_1+80\theta_2+18$	0.24375	0.05625	29	29
	P2	0.2625	0.0375	28	28					
	P3	0.2625	0.075	31	31					
	P4	0.225	0.075	30	31					
CR37	P1	0.225	0.075	30	30	$T=\frac{40}{3}\theta_1+\frac{200}{3}\theta_2+22$	0.2625	0.1125	33	33
	P2	0.3	0.075	31	31					
	P3	0.3	0.15	36	36					
	P4	0.225	0.15	35	35					
CR38	P1	0.15	0.075	28	28	$T=\frac{80}{3}\theta_1+\frac{200}{3}\theta_2+19$	0.1875	0.1125	32	32
	P2	0.225	0.075	30	30					
	P3	0.225	0.15	35	35					
	P4	0.15	0.15	33	33					
CR39	P1	0.15	0.15	33	33	$T=20\theta_1+60\theta_2+21.25$	0.1875	0.1875	36	36
	P2	0.225	0.15	35	35					
	P3	0.225	0.225	39	39					
	P4	0.15	0.225	38	38					
CR40	P1	0.225	0.15	35	35	$T=\frac{40}{3}\theta_1+\frac{160}{3}\theta_2+24$	0.2625	0.1875	38	38
	P2	0.3	0.15	36	36					
	P3	0.3	0.225	40	40					
	P4	0.225	0.225	39	39					
CR41	P1	0.225	0.225	39	39	$T=\frac{40}{3}\theta_1+40\theta_2+27$	0.2625	0.2625	41	41
	P2	0.3	0.225	40	40					
	P3	0.3	0.3	43	43					
	P4	0.225	0.3	42	42					
CR42	P1	0.15	0.225	38	38	$T=\frac{40}{3}\theta_1+40\theta_2+27$	0.1875	0.2625	40	40
	P2	0.225	0.225	39	39					
	P3	0.225	0.3	42	42					
	P4	0.15	0.3	41	41					
CR43	P1	0	0.15	30	30	$T=\frac{80}{3}\theta_1+\frac{200}{3}\theta_2+20$	0.0375	0.1875	34	34
	P2	0.075	0.15	32	32					
	P3	0.075	0.225	37	37					
	P4	0	0.225	35	35					
CR44	P1	0.075	0.15	32	32	$T=\frac{40}{3}\theta_1+\frac{200}{3}\theta_2+21$	0.1125	0.1875	35	35
	P2	0.15	0.15	33	33					
	P3	0.15	0.225	38	38					
	P4	0.075	0.225	37	37					
CR45	P1	0.075	0.225	37	37	$T=\frac{40}{3}\theta_1+40\theta_2+27$	0.1125	0.2625	39	39
	P2	0.15	0.225	38	38					
	P3	0.15	0.3	41	41					
	P4	0.075	0.3	40	40					
CR46	P1	0	0.225	35	35	$T=20\theta_1+\frac{140}{3}\theta_2+24.75$	0.0375	0.2625	38	38
	P2	0.075	0.225	37	37					
	P3	0.075	0.3	40	40					
	P4	0	0.3	39	39					

In this case study, two parameters, the energy loss for coal and electricity in city A, are estimated before the energy dispatch optimization. They are actually typical uncertainties that should be investigated to disclose their impact to the final energy dispatch optimization results. Assume they can vary within the range from 0 (no energy loss) to 0.3 (30% energy loss). Then based on the proposed algorithm of uncertainty space partition, a two-dimensional uncertainty space has been constructed, where uncertainties θ_1 represents the energy loss fraction of coal ($0 \leq \theta_1 \leq 0.3$) and θ_2 represents the energy loss fraction of electricity ($0 \leq \theta_1 \leq 0.3$). Figures 19 and 20 show the partitioning result with the total of 46 critical regions (CR1~CR46). The detailed regressed functions for the optimal dispatch time with respect to θ_1 and θ_2 as well as the validation results are shown in Table 2, where the standard criteria (SC) of derivative percentage is set as 1.5% to identify the approximation performance of function regression.

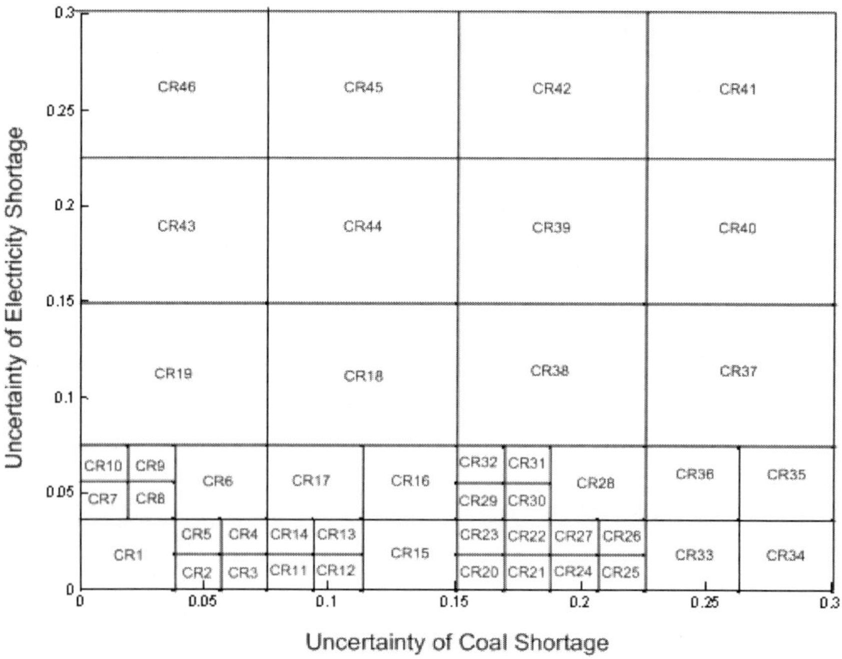

Figure 19. Uncertainty space partitioning for the case study.

Note that the 3D profile in Figure 20 is not a smooth surface although it looks like. Together with Table 2, they characterize the relation between the optimal energy dispatch time and the studied two uncertain parameters. They provide decision makers a reference manual for the easy lookup of the possible impact from uncertainties and thus help conduct right decisions. For example, the 3D surface generally shows the less coal and electricity shortages will result in less energy recovery time. As its 2D projection on the uncertainty space as shown in Figure 21, any emergency case specified by coal and electricity shortage for City A corresponds to a point (say point S) in the plane.

The energy dispatch behaviors among the network are actually moving the point S to finally reach the left-bottom corner point. It should be known that any movement in a short time period can be decomposed into two directions: one is toward the iso-time direction that does not reduce energy recovery time at all; the other is gradient direction perpendicular to the iso-time direction, which represents the shortest energy recovery time and smart move direction. For illustration, the iso-time and gradient directions at a point S (with 20% electricity and 15% coal shortages respectively) are demonstrated in Figure 21. Based on such information, the dynamic impact of different energy dispatch strategies can be evaluated, which will help decision makings under uncertainties.

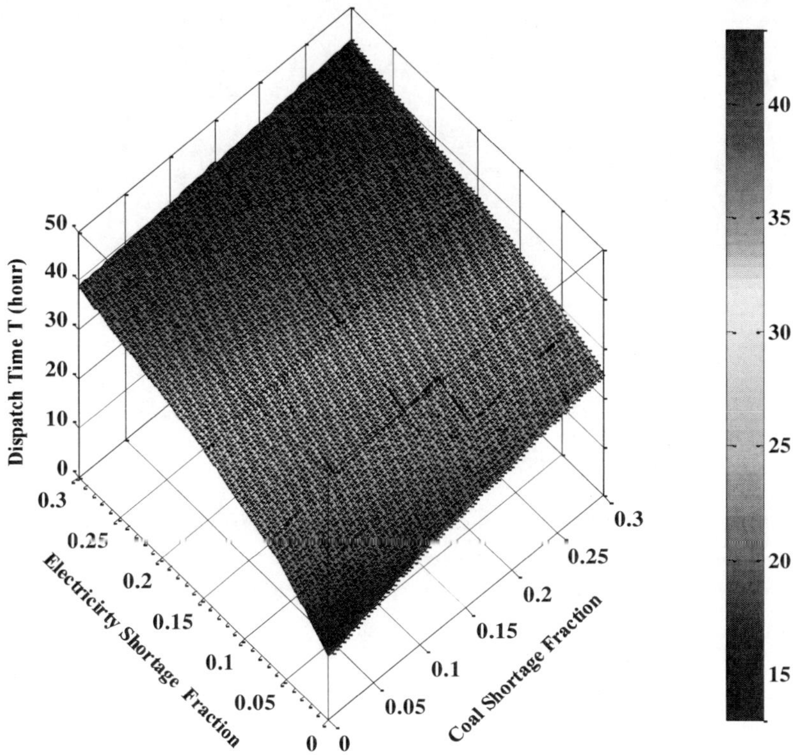

Figure 20. Minimum energy dispatch time with respect to uncertainties.

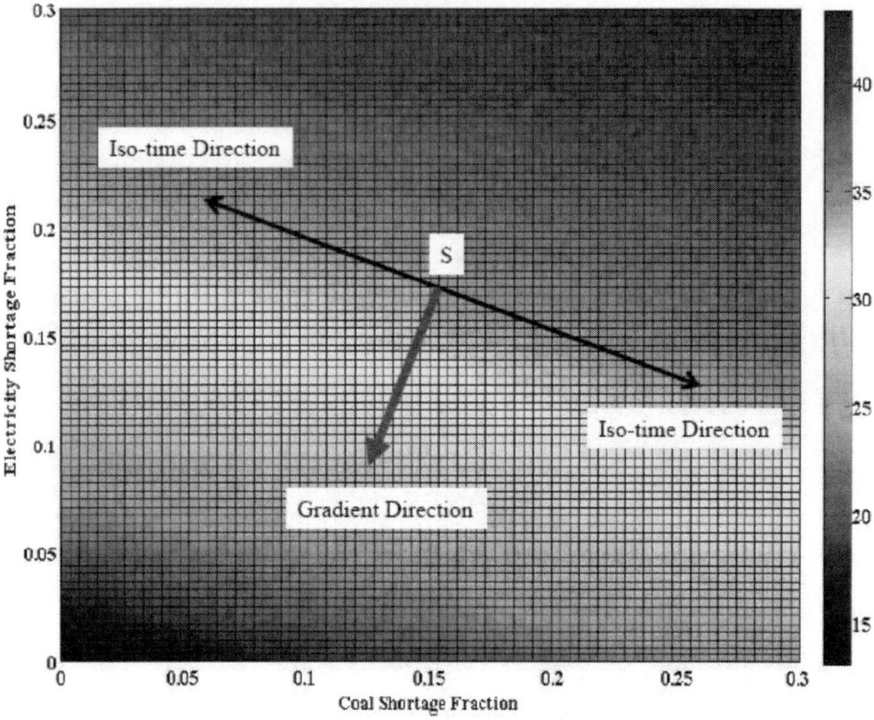

Figure 21. Application analysis based on uncertainty space partitioning.

CONCLUSION AND FUTURE WORK

Numerical modeling or mathematical modeling has become an essential tool in most branches of science and engineering. It aims to handle the process of solving physical problems by appropriate simulation of reality. With the increase in computational technology, many numerical models and software programs have been developed for various engineering practices. The currently developed tsunami modeling tools has provided numerical models for faster and more reliable forecasts of tsunamis propagating through the ocean and striking coastal communities but it should still be noted that we are lack of the numerical modeling to take the responsibility of decision making for emergency response planning such as infreastructure recovery and energy nework dispatch after the damage of tsunami. Although it is based on some simplifications and assumptions, our developed MILP based model for energy network dispatch optimization under emergency of local energy shortage after the natural disaster such as tsunami is the first prototype in the area of energy dispatch optimization under emergencies. The sensitivity analysis method is also conducted with respect to uncertainties by partitioning the entire space of uncertainty parameters into multiple subspaces. Hope our preliminary study can become the starting point for the future development and real application of mathematical modeling in response planning under emergency of tsunami, expecially for dispatch optimization in agile energy network.

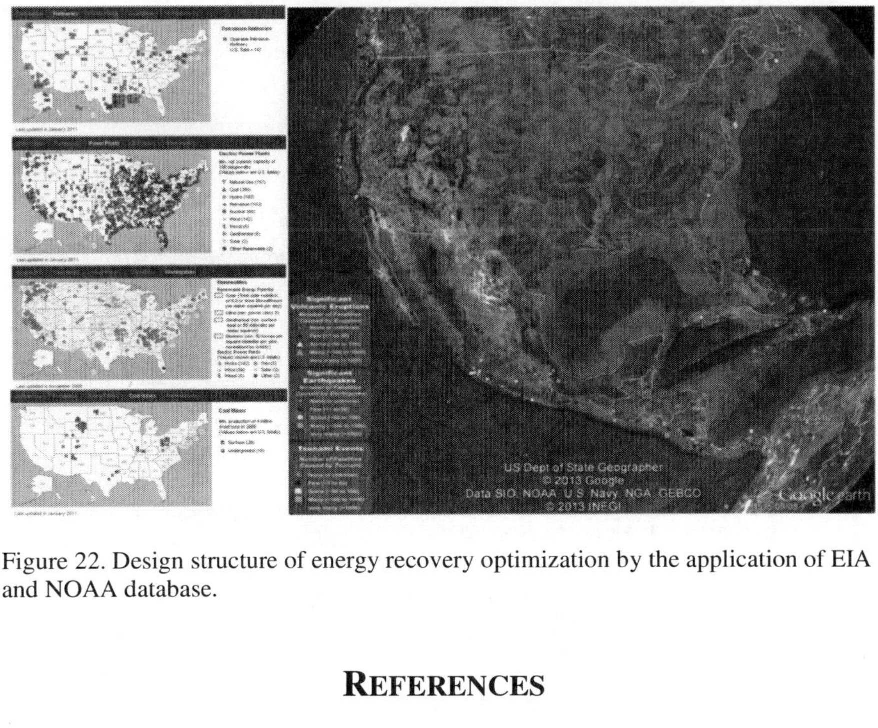

Figure 22. Design structure of energy recovery optimization by the application of EIA and NOAA database.

REFERENCES

[1] Fradin, J. B. a. D. B., 2008. *Witness to Disaster:Tsunamis.* Washington,D.C.: National Geographic Society.

[2] Bernard, E.N. (1999): *Tsunami. Natural Disaster Management*, Tudor Rose, Leicester, England, 58-60.

[3] Lavelle, M., 2012. National Geographic. [Online] Available at:*http://news. nationalgeographic. com /news/energy/2012/03/120309-japan-fukushima-anniversary-energy-shortage/* [Accessed 16 December 2012].

[4] Bernard, E.N. (1998): Program aims to reduce impact of tsunamis on Pacific states. *Eos Trans. AGU*, 79(22), 258, 262-263.
[5] Synolakis, C., P. Liu, G. Carrier, H. Yeh, *Tsunamigenic Sea-Floor Deformations, Science*, 278, 598-600, 1997.
[6] Dudley, Walter C., and Min Lee (1998): *Tsunami! Second Edition,* University of Hawai'i Press, Honolulu, Hawaii.
[7] Cornell Multi-grid Coupled TsunamiModel, *http://ceeserver.cee.cornell.edu / pll-group/r2comcot.html*
[8] Liu, P. L. F., Cho, S., Yoon, S. B., and Seo, S. N.: Numerical simulations of the 1960 Chilean tsunami propagation and inundation at Hilo, Hawaii, Recent Development in Tsunami Research, in: Tsunami: progress in prediction, disaster prevention and warning, edited by: Tsuchiya, Y. and Shuto, N., *Adv. Nat. Technol. Haz.*, 99–115, Springer, 1994.
[9] Liu, P. L.-F., Cho, Y.-S., Briggs, M. J., Kanoglu, U., and Synolakis, C. E.: Runup of solitary waves on a circular island, *J. Fluid Mech.,* 302, 259–285, 1995.
[10] Wang, X. and Liu, P. L.-F.: A Numerical investigation of Boumerdes-Zemmouri (Algeria) Earthquake and Tsunami, Comp. *Model Eng.*, 10, 171–184, 2005.
[11] Wang, X. and Liu, P. L.-F.: An analysis of 2004 Sumatra earthquake fault plane mechanisms and Indian Ocean tsunami, *J. Hydraul. Res.*, 44, 147–154, 2006.
[12] PDC, *http://www.pdc.org/iweb/tsunami_modeling.jsp*
[13] NOAA Tsunami, *http://www.tsunami.noaa.gov/index.html*
[14] Tsunami Computational Portal, *http://tsunamiportal.nacse.org/*
[15] The Global Sea Level Observing System, *http://www.gloss-sealevel.org/*
[16] Bellarmine GT, Arokiaswamv NSS. Energy Management Techniques to Meet Power Shortage Problems in India. *Energy Conversion & Management*. 1996; 37(3):319-328.
[17] Meier A. Operating Building During Temporary Electricity Shortage. *Energy & Buildings*. 2006; 38(11):1296-1301.
[18] Sanghvi AP. Power Shortages in Developing Countries: Impacts and Policy Implications. *Energy Policy*. 1991; 19(5): 425-440.
[19] Heffner G, Maurer L, Sarkar A, Wang X. Minding the Gap: World Bank's Assistance to Power Shortage Mitigation in the Developing World. *Energy*. 2010; 35(4): 1584-1591.

[20] Kainkwa RR. Wind Energy as An Alternative Source to Alleviate The Shortage of Electricity That Prevails during the Dry Season: A Case Study of Tanzania. *Renewable Energy*. 1999; 18(2):167-174.

[21] Berredo RC, Ekel PY, Martini JSC, Palhares RM, Parreiras RO, Pereira JG. Decision Making in Fuzzy Environment and Multi-Criteria Power Engineering Problems. *International Journal of Electrical Power & Energy Systems*. 2011; 33(3):623-632.

[22] Gal T, Nedoma J. Multiparametric Linear Programming. *Manage. Sci.* 1972; 18:406–422.

[23] Acevedo J, Pistikopoulos EN. An Algorithm for Multiparametric Mixed-integer Linear Programming. *Oper. Res. Letters.* 1999: 24:139–148.

[24] Dua V, Bozinis NA, Pistikopoulos EN. A Multiparametric Programming Approach for Mixed-integer Quadratic Engineering Problems. *Comput. Chem. Eng.* 2002; 26:715–733.

[25] Acevedo J, Salgueiro M. An Efficient Algorithm for Convex Multiparametric Nonlinear Programming Problems. *Ind. Eng. Chem. Res.* 2003; 42:5883–5890.

[26] Acevedo J, Pistikopoulos EN. A Hybrid Parametric/stochastic Programming Approach for Mixed-integer Linear Problems under Uncertainty. *Ind. & Eng. Chem. Research.* 1997; 36(6): 2262-2270.

[27] Acevedo J; Pistikopoulos EN. A Multiparametric Programming Approach for Linear Process Engineering Problems under Uncertainty. *Ind. & Eng. Chem. Research.* 1997; 36(3): 717-728.

[28] Ryu JH, Dua V, Pistikopoulos EN. A Bilevel Programming Framework for Enterprise-wide Process Networks under Uncertainty. *Computers & Chemical Engineering*. 2004; 28:1121-1129.

[29] Jia Z, Ierapetritou MG. Generate Pareto Optimal Solutions of Scheduling Problems Using Normal Boundary Intersection Technique. *Computers & Chemical Engineering*. 2006; 31:268-280.

[30] Jia Z, Ierapetritou MG. Uncertainty Analysis on the Right-hand-side for MILP Problems. *AIChE J.* 2006; 52:2486-2495.

[31] Verderame PM, Floudas CA. Operational Planning of Large-scale Industrial Batch Plants under Demand Due Date and Amount Uncertainty: I. Robust Optimization. *Ind. & Eng. Chem. Research.* 2009; 48:7214-7231.

[32] *CPLEX. Using the CPLEX callable library*. CPLEX Optimization, Inc.: Incline Village, NV. 2009.

[33] GAMS. *GAMS- A User' s Guide.* GAMS Development Corporation: Washington, DC. 2009.

INDEX

D

E

F

G

H

I

N

O

P

Q

R

S